The State, Popular Mobilisation and Gold Mining in Mongolia

ECONOMIC EXPOSURES IN ASIA

Series Editor:
Rebecca M. Empson, Department of Anthropology, UCL

Economic change in Asia often exceeds received models and expectations, leading to unexpected outcomes and experiences of rapid growth and sudden decline. This series seeks to capture this diversity. It places an emphasis on how people engage with volatility and flux as an omnipresent characteristic of life, and not necessarily as a passing phase. Shedding light on economic and political futures in the making, it also draws attention to the diverse ethical projects and strategies that flourish in such spaces of change.

The series publishes monographs and edited volumes that engage from a theoretical perspective with this new era of economic flux, exploring how current transformations come to shape and are being shaped by people in particular ways.

The State, Popular Mobilisation and Gold Mining in Mongolia

Shaping 'Neoliberal' Policies

Dulam Bumochir

First published in 2020 by
UCL Press
University College London
Gower Street
London WC1E 6BT

Available to download free: www.uclpress.co.uk

© Dulam Bumochir, 2020

The author and editor has asserted his rights under the Copyright, Designs and Patents Act 1988 to be identified as the author of this work.

A CIP catalogue record for this book is available from The British Library. This book is published under a Creative Commons 4.0 International licence (CC BY 4.0 International). This licence allows you to share, copy, distribute and transmit the work; to adapt the work and to make commercial use of the work providing attribution is made to the authors (but not in any way that suggests that they endorse you or your use of the work). Attribution should include the following information:

Bumochir, Dulam. 2020. *The State, Popular Mobilisation and Gold Mining in Mongolia: Shaping 'Neoliberal' Policies*. London: UCL Press. DOI: https://doi.org/10.14324/111.9781787351837

Further details about Creative Commons licences are available at http://creativecommons.org/licenses/

Any third-party material in this book is published under the book's Creative Commons licence unless indicated otherwise in the credit line to the material. If you would like to re-use any third-party material not covered by the book's Creative Commons licence, you will need to obtain permission directly from the copyright holder.

ISBN: 978-1-78735-185-1 (Hbk)
ISBN: 978-1-78735-184-4 (Pbk)
ISBN: 978-1-78735-183-7 (PDF)
ISBN: 978-1-78735-186-8 (epub)
ISBN: 978-1-78735-187-5 (mobi)
DOI: https://doi.org/10.14324/111.9781787351837

Contents

List of acronyms	vi
Note on transliterations	vii
Acknowledgements	viii
Preface	xi
Introduction	1
1. The Reification of the National Economy	21
2. Beyond 'Resource Nationalism'	40
3. Navigating Nationalist and Statist Initiatives	60
4. Advocacy and Activism in Popular Mobilisations	81
5. The De-deification of the State	101
6. An Original Environmentalist Society	121
Conclusion	136
References	142
Index	155

List of acronyms

ADB	Asian Development Bank
CGM	Cold Gold Mongolia
ERC	European Research Council
FDI	Foreign direct investment
FN	Fire Nation
HWPC	Homeland and Water Protection Coalition
IMF	International Monetary Fund
GDAT	Group for Debates in Anthropological Theory
KAS	Konrad-Adenauer-Stiftung
MDP	Mongolian Democratic Party
MNMA	Mongolian National Mining Association
MNPC	Mongolian Nature Protection Coalition
MPR	Mongolian People's Republic
MPRP	Mongolian People's Revolutionary Party
MRPAM	Mineral Resources and Petroleum Authority of Mongolia
NUM	National University of Mongolia
ORM	Ongi River Movement
OT	Oyu Tolgoi
PRC	People's Republic of China
TAF	The Asia Foundation
TNCs	Transnational corporations
UCL	University College London
UMMRL	United Movement of Mongolian Rivers and Lakes
US	United States
WB	World Bank

Note on transliterations

For Mongolian Cyrillic transliteration, I adopted the MNS 5217:2012 standard approved by the National Council of Standardisation (*Standartchillyn ündesnii zövlol* 2012) of Mongolia. For Mongolian names of people, however, I had to follow the system General Authority for State Registration of Mongolia adopted to print names on Mongolian passports, which is the widely accepted system to write contemporary Mongolian names. The main difference between the MNS 5217:2012 standard and the registration authority standard appear in Mongolian Cyrillic letters such as Θ, Y and У. In the state registration authority standard, the transliteration of the Mongolian Cyrillic Θ is U which is not different from the transliteration of the Mongolian Cyrillic Y and У. In the MNS 5217:2012 standard, the transliteration of the Mongolian Cyrillic Θ is Ö, while the transliteration of the Mongolian Cyrillic Y is Ü and У is U. For example, the name 'Мөнх' is 'Munkh', not 'Mönkh', in the state registration authority standard which I adopt in this book. Also, in the case of some well-known Mongolian names and words, I followed the most widely accepted versions in English literature. For example, 'Буриад', which should be 'Buriad' in the MNS 5217:2012 standard, often appears as 'Buryat' in English literature. For Mongolian classical script, known as Uyghur, Vertical or Old Mongolian script, I adopted B. I. Vladimirtsov's (1971) system, except Q replaces KH, SH replaces Š, and GH replaces γ in some words that are already established.

Acknowledgements

I could not have written this book without the generosity, compassion, affection and trust of the many people who supported and shaped not only this book but also the period of my professional and personal life during the time it took to research and write it. First, I sincerely appreciate those who trusted and shared their life experiences, feelings, faith and hope during my fieldwork in Mongolia, and permitted me to write about them and their lives. Many of the prominent figures that I depict in this book – in the protest movements, in mining and in politics – have differing relationships that can lead to disagreements and conflicts, and they expect reactions of different readers who might critique, judge and blame them; still, all of them kindly allowed me to depict them here. For this reason, I ask readers to respect the actions, decisions and beliefs of the different agents and actors who will appear in this book.

One of those prominent figures is Munkhbayar Tsetsegee, a founder and leader of the river movements. Together with his family, he welcomed me and my colleague Byambabaatar Ichinkhorloo to stay at his home during my field trips in Saikhan-Ovoo, Dundgovi, in the south of Mongolia. I also thank Bayarsaikhan Namsrai, another prominent founder of the movement, and Dashdemberel Ganbold, a lawyer of the movement, for providing invaluable information and research materials. The time I spent with these people in the movement helped me to discover the true meanings and realities behind the tag 'green terrorists'. I was also able to uncover their concerns and relentless fight to protect the environment and lives of local people.

I received invaluable help from some people in the mining industry, in order to depict the other side of the story. I sincerely thank Orgilmaa Zundui-Yondon and Danny Walker for permitting me to conduct field research at their gold mine site in Zaamar, in the central region of Mongolia. I was impressed with their advanced mining and rehabilitation technology, as well as their sincere efforts, time and deeds that they have dedicated to rehabilitate the environment. The same also applies to Chinbat Lhagva, a former geologist and miner, founder of Gantsuurt company, and his daughter (and my friend) Nomin Chinbat, and her husband (and my friend) Bat-Erdene Gankhuyag. I am grateful for their help as I tried to educate myself in the fields of geology and mining.

In addition to those involved in the river movements and the mining industry, many other people helped shape this work. These include many technocrats and politicians, namely, Algaa Namgar, a metallurgical engineer and a director of the Mongolian National Mining Association (MNMA), and Jargalsaikhan Dugar, a mining economist and a former director of MNMA, who provided inestimable research materials and helped me to conduct field research. I am grateful to Algaa and Jargalsaikhan for helping me to meet and interview other people who had important political positions in the ruling institution of Mongolia – those are Ochirbat Punsalmaa, a former president of Mongolia and a mining engineer, and Byambasuren Dash, a former prime minister and economist who ruled Mongolia in the 1990s. Their insights about the rule of the country after the collapse of socialism made a significant contribution as I worked to write about political independence, geopolitics and national economy.

I would like to express my deepest gratitude and warm feelings of indebtedness to my *bagsh* (masters), colleagues and friends. Those are my *bagsh* Caroline Humphrey and *bagsh* David Sneath at the University of Cambridge, who mentored, inspired and supported my scholarly works throughout my academic career (including my time as a student). I am also grateful to Munkh-Erdene Lhamsuren at the National University of Mongolia (NUM), Uradyn Bulag at the University of Cambridge and Morten Pedersen at the University of Copenhagen for sharing thoughts and advice, which helped me to realise my oversights and improve my arguments.

I am also deeply indebted and endlessly appreciative to my colleagues in the University College London (UCL) team of the Emerging Subjects research project, funded by the European Research Council (ERC). I am grateful to Rebecca Empson, a principal investigator of the project, whose expert knowledge, excellent academic skills and intellectual visions granted an opportunity for me to hold a research associate post at UCL and to produce this book. Lauren Bonilla and Rebekah Plueckhahn, the two other research associates of the project; Hedwig

Waters, a doctoral candidate of the project, and Elisabeth Fox and Joseph Bristley, the two other doctoral candidates – I have spent extraordinarily productive and fascinating time with all of you in the office at UCL and the field in Mongolia from 2015 to 2019. Thank you so much for being there throughout the research project and the writing up of this book, and for honestly critiquing, commenting, advising and helping to polish and improve my writing.

I also genuinely appreciate the help of my colleague Byambabaatar Ichinkhorloo, who assisted my fieldwork in Mongolia, and Michelle Beckett and Joe Ellis, who helped me to edit and revise this book. Finally, this book was written during research funded by the H2020 European Research Council [ERC-2013-CoG, 615785, Emerging Subjects], and I thank the ERC for providing such a fantastic opportunity.

Preface

Mongolia is known to the world for its post-Soviet democratic and neoliberal transformation, successfully achieved in complete peace, without any fighting or violence. However, more than 20 years after the collapse of the socialist regime, environmental and nationalist protestors interrupted the peace in an attempt to gain attention for the protection of the environment from mining and to preserve the mobile pastoral way of life. On 16 September 2013, there was a gunshot[1] during a protest carried out by several environmentalist and nationalist movements in Chinggis Square[2] in front of the state house (*Töriin ordon*), where the president, parliament and government of Mongolia operate. Mongolia's most famous and most successful environmental activist, Goldman Environmental Prize winner (2007) and National Geographic Emerging Explorer (2008) Munkhbayar Tsetsegee, appeared in front of the state house entrance with his colleagues, armed with rifles and grenades. State special security (*Töriin tusgai khamgaalaltyn gazar*), intelligence (*Tagnuulyn yerönkhii gazar*) and police terminated the actions of the protestors and arrested them. No one was hurt. Munkhbayar was arrested, along with 11 men of the river movement. The court sentenced five members of the movement to one to ten years in prison for possessing arms, attempted terrorism and extortion (see Chapter 5). The incident shocked many Mongolians because it was probably the first ever public incident, in modern times, of Mongolians bringing arms against the *tör* (state),[3] which is considered as superior, sacred, respected and unchallengeable by many Mongolians (see also Bumochir 2004; Dulam 2009). The destructive consequences of the mining industry generated seemingly endless protests by environmentalist and nationalist movements, and some activists professed that they were willing to sacrifice their own lives to protect the environment. Amongst many other motivations, the above incident inspired me the most to write this book and to understand what made these activists decide to make such a move.

Since 1990, as a Mongolian I grew up learning to envision the future of Mongolia as a modern nation like Japan, where preservation of *ulamjlal* (tradition) and advancement of *shinechlel* (modernisation) can happen together. Yet for many Mongolians a combination of the two sometimes appears to be a dilemma. In other words, my experience with the topic of the book is itself a dilemma. For example, taking the difficulties of the economy and the environment into consideration, many Mongolians think that our nation-state and its people are facing a dilemma (see also Zulbayar 2015; Jargalsaikhan 2018). The dilemma is between modern and traditional, local and global, nomadic and urban, pastoralism and mining, socialism and capitalism, nationalism and neoliberalism, and the environment and economy. Mongolians who experience the extreme consequences of the economic boom and bust, and witness endless political debates between neoliberal and nationalist agents, seem to be torn between the contesting ideologies that rule the country (see also Munkherdene 2018). For example, at the end of 2014, when Mongolian Prime Minister Saikhanbileg Chimed (2014–16) established his cabinet, he immediately announced that the country was encountering a severe economic crisis. He was enthusiastic to urgently start large mining projects and appeal to investors as soon as possible to aid the declining economy (Dulam 2015; Bumochir 2017, 30; Odonchimeg 2015). One of the projects was the Gatsuurt gold mine, on a historical sacred mountain named Noyon, co-owned by Canadian Centerra Gold Inc. and the Mongolian state. In January 2016, in the parliamentary session to approve the government's decision to contract Canadian Centerra Gold to extract at the sacred mountain – with dozens of invaluable archaeological sites from the Xiongnu empire (3rd BC–1st AD) – Saikhanbileg responded to the dissent of parliament member, former journalist and activist Uyanga Gantumur[4] with agony and frustration:

> Now we have just taken Noyon Mountain under protection [historical site]. In ten kilometres' distance, there is the Gatsuurt deposit, which was left halfway extracted and we are about to continue and finish [the extraction]. As a result, about 1,000 jobs will be available, US$240 million – which is MNT 480 billion – will be our tax. Plus, we will have 50 tonnes of gold reserve. These 50 tonnes of gold reserve will help the dollar rate you are paying for

from your pocket. Therefore, we are doing this in order to bring our economy into circulation. You demand the government to improve the economy, and when the government tries to do something, then you demand the government resign. In your action of a pair of scissors, what kind of government, what kind of state and what kind of Mongolia can move forward and develop?[5]

In parliament, Uyanga actively represented the voices of nationalist and environmentalist movements and fought against extractive industry destruction and mining corporations. In that sense, this was a reaction not only to Uyanga and other members in parliament who resisted but also to vast numbers of environmental and nationalist movement members, local residents and the public, who campaigned and protested for more than a year by organising hunger strikes, performing shamanic rituals on the central square, worshipping the mountain, and developing scholarly documentation of archaeological findings. The parliamentary session issued a decision approving the government proposal to register 11,000 hectares of the Noyon Mountain in the network of special state protection (*töriin tusgai khamgaalalt*), under the category of 'natural resource', and 405 hectares in another network of state special protection, under the category of 'historical site' (also singularity). Parliament also approved the government's deal with Canadian Centerra Gold, and agreed Mongolian state ownership to be 34 per cent. These two categories of the state protection of the same mountain divide the mountain into two parts. The smaller part, designated a 'historical site', is to protect the sacred mountain, as some protestors and politicians in parliament demanded. The larger part is a 'natural resource' to mine the gold deposit – with the investment of Canadian Centerra Gold company – in order to ease Mongolia's economic crisis.

As we saw above, the prime minister described contradictory necessities and demands of actors as two blades of a pair of scissors. This is an interesting way to present the situation of the government in a dilemma of two conflicting goals: to secure the economic development of the country and also to protect its history and environment.[6] With the metaphor of the scissors, if the government promotes one of the necessities or demands and drops the other, then the other necessity (depicted as the other scissor blade) cuts or causes damage. Therefore, to keep both sides happy, the government decided to accept and promote both of the contradicting demands at the same time. The prime minister's presentation of the situation is indicative of his desperation to solve the dilemma. However, there are those who do not accept the decision, and they argue that the scissor blade represents the economy/ development and that corporations will still cause damage by scissoring historical and natural sites. As a reaction, the other scissor blade represents the nationalist movements and environmental attempts to scissor the government, corporations and the project to aid the national economy. Therefore, the prime minister had to actively demand Uyanga and the other politicians and movements – representing the opposition, the other blade of the scissors – to cease their resistance. In the prime minister's description, the movements are depicted as being in control of one of the two blades of a pair of scissors. The movement can still inflict damage, even if the government manages to secure the action of the other scissor blade. The government does not have the power or authority to fully control the situation, as it can secure only one of the scissor blades, not two.

For those who did not accept the above decision of the government, the first dilemma of economy/development versus history/environment is false. They argue that there are other ways to solve the crisis of the economy – that is, rather than by extracting the sacred mountain. The prime minister justifies his decision by explaining that this was the best solution to save the national economy and the environment at the same time. However, it is difficult to know whether the decision was solely for the sake of the national economy or for the environment. How much consideration was given to his individual reputation and the interest of his political party? The prime minister wanted to keep and increase the number of supporters' votes in the coming parliamentary election, which was less than six months away. People also suspect that behind the discourse of scissors and dilemmas, there exists corruption, individual deals with investors and donations to political parties. In other words, the presentation of the political and economic situation in the framework of a dilemma is a useful tool to obscure reality and justify political decisions.[7]

However, many also question whether the economic and environmental dilemmas are inevitable. For example, Ian Goldin and Alan Winters (1995, 14) conclude that 'economic growth and development are perfectly consistent with environmental protection'. While others, namely Herman E. Daly (1996, 1), suggest that terms such as the economy and the environment together create an 'oxymoron'. Moreover, Raúl R. Cordero, Pedro Roth and Luis Da Silva (Cordero et al. 2005, 1) conclude that environmental care and economic growth are not incompatible but to reconcile them is not easy.[8] Paul Ekins (2002), in his account of 'green growth', writes about how to find the compatibility between economic growth and environmental sustainability. While it is not easy, it is possible, and thus they consider it to be a false dilemma.

I do not aim to contribute to this debate and prove or disprove contradiction or consistency in the relationship of the economy and the environment. Instead, it is interesting to see how the resource economy and environment are believed to be a dilemma in certain cultures, or how the so-called dilemma develops into a conflict, or how a false dilemma is presented to be a real dilemma for political purposes. These questions about dilemmas are useful to understand and show the complexity of the multipartite relationship of agents in the state, mining companies, donors and movements. In other words, despite what international financial and development organisations suggest – and what scholars debate – true or false, some Mongolian rulers present certain challenges as a dilemma or struggle, and fail to find the difficult or perhaps non-achievable marriage of economic growth *and* environmental protection.

In his discussion about the 'resource trap', Paul Collier notes that 'each rich, resource-hungry country is locked into a prisoner's dilemma of inaction' (Collier 2007, 47). Similarly, Stuart Kirsch also notes that deep in the heart of the resource problem there are 'underlying dilemmas associated with the capitalist modes of production [that] can never be completely resolved; they can only be renegotiated in new forms' (Kirsch 2014, 3). In other words, it is a mechanism of neoliberal capitalism, foreign investment and forms of financialisation that forces the state into dilemmas. For Collier, the dilemma imprisons and prevents action, while for Kirsch, the dilemma cannot be resolved entirely, but it can be renegotiated. Renegotiation is precisely what the Mongolian government had been attempting.

It is useful to take this brief account of Collier and Kirsch into consideration and explore experiences of dilemmas as they emerged in Mongolia. For example, Katherine Verdery and Caroline Humphrey (2004, 17) discuss the postsocialist dilemma or 'political impasse' and show how the government of Mongolia tangled with the issue of whether (and how) to acknowledge the Mongols' cultural practices of past eras and international advice to create private property. In reference to natural resources and mining, Mette High (2012, 249) describes how 'some historical epochs have sought to limit the extraction of minerals in the Mongolian cultural region, others have celebrated mining for its potential promotion of economic growth and large-scale industrialisation' (see also High and Schlesinger 2010). In their critique against the accusation of 'resource nationalism', Rebecca Empson and Tristan Webb (2014, 232) argue that the Mongolian state[9] attempts to establish 'trusting partnerships'. 'The idea of "trusting partnerships" can refer to the relationship between the Mongolian State and foreign investors, as well as in specific ways to that relationship between the State and the Mongolian people'. In this relationship, the state struggles to 'balance expectations from all of the partnerships' (Empson and Webb 2014, 247).

In this book I present the different dilemmas of the national government, parliament and rulers of Mongolia caught between liberal, neoliberal, market and capitalist, and the so-called populist, nationalist and patriotic tendencies and approaches: the establishment of the liberal economy and mining industry in Chapter 1; state control in Chapter 2; and the approval of the environmental protection law in Chapter 5, which were all political 'renegotiations' made under the pressure of different dilemmas between environmental well-being and economic prosperity, and popular mobilisations and mining corporations and investors. Anticipating the political, economic, social, cultural, religious and environmental dilemmas of the nation-state government, rulers, local residents and nationalist movements in the neoliberal, democratic and global world is an alternative way to understand and interpret difficulties of countries such as Mongolia. Pascale Hatcher (2014, 128) argues that 'the cases of the Philippines and Mongolia, and to a lesser extent Laos, rather show that the recent changes in policy are symptoms of the increasing dilemma forced upon the state by the very third generation of mining regimes promoted by the multilateral institutions'. In the sense of being a force produced by multilateral institutions or agents, false or not, dilemmas tell us about struggles, challenges of individual rulers, protestors, local residents, company owners and investors and all other inner workings in the resource-abundant country.

As a Mongolian who shares experiences of such dilemmas, in order to draw a comprehensive picture of Mongolian resource economy, environment, mobilisation, nationalism and state, I conducted research on multilateral institutions and agents. To do this, I met and interviewed political leaders such as Ochirbat Punsalmaa, the first president of Mongolia (1990–7), and Byambasuren Dash, the last prime minister of the Mongolian People's Republic (1990–2), and their colleagues and other politicians and technocrats, who had prominent roles in the establishment of the liberal mining economy. In addition to such politicians and officials, I also visited Mongolian and foreign gold mining companies and interviewed their owners, operators and managers, who support the liberal economy and free market principles. They directly benefited from the liberal policy on mining economy; they were also disturbed by the results of laws, regulations and political decisions to control natural resources as well as successful protests of nationalist environmental movements. In order to provide an account of the other side of the conflict, I met the so-called 'resource nationalist' politicians, economists, lawyers, technocrats and scholars, who succeeded in implementing different forms of state control over natural resources and placed restraints on mining companies. The popular mobilisations make up another influential group of people who contributed to nationalistic politics. Local people – including herders and those who inhabit the administrative unit centre settlements – are those who were most negatively affected by the destruction of the mining operations and started fighting against gold mining companies and the state. They are also the ones who contributed most to shape nationalism, environmentalism and the state in Mongolia: they closed down dozens of mining operations and stalled some hundreds of mining licences, and defeated the government in the supreme court when it did not implement a law to protect the environment.

As an anthropologist, I sought to understand and describe the above-mentioned opposing groups and their approaches. I informed and explained to everyone I met that I would also meet and interview people who were in direct opposition and that this was important to draw a full picture of the scenario. All of the people I met understood my situation and accepted my position as an anthropologist and a Mongolian who was trying to develop a multi-faceted approach. I ended up in relationships with these people – I worked with many people with competing ideologies for years – and I know that I cannot judge them or prioritise any of them over any other. It is impossible (or futile) to prove or disprove all of the information that emerged in the interviews – for example, I cannot prove or disprove popular suspicions about nationalist movements and the extortion of mining companies; I cannot confirm or deny the corruption of politicians; and there is no final word to be found regarding private mining company owners who possess many mining licences, which make up a significant portion of Mongolian mineral wealth. In my work, I present my materials as they present themselves, and how these actors make sense of their approaches, and how they justify their actions. This book does not prove or disprove whether the liberal or nationalistic political decisions and actions were right or wrong. Instead, it lets opposing voices be heard, and allows them to propose their justifications.

My position as an anthropologist is not a dilemma for me, because I can write about all of these people, about both neoliberalism and nationalism and how they conflict and resist each other. However, my personal position between the two ideologies and those who promote them is a dilemma for me; that is, as a Mongolian. Here, my dilemma should not be confused with my anthropological concerns with ethics, methodology and the scholarly discourses with which my work engages. However, as a Mongolian intellectual, I do worry about the country's vulnerable condition, just as many other Mongolians do. Many Mongolians ask me whether the neoliberal or nationalistic approaches are correct for Mongolia to advance. This might be what many readers seek an answer to when they read this book. After observing both sides, I concluded that contemporary Mongolia reveals the interaction of indigenisation and neoliberalisation, and how different dilemmas force the two processes to shape one another. For example, a dilemma between the economy and the environment – real or not – forces the indigenous aspects of the nation to shape neoliberal policies and the capitalist free market. In other words, the indigenous shaping of neoliberalism is a product of different dilemmas. In the same way, I find that in consequence of a dilemma between neoliberalism and nationalism, or the economy and environment in Mongolia, matters of the indigenous nation-state shape neoliberal policies and markets. This should be expected. As Amarjargal Rinchinnyam, the former prime minister of Mongolia (1999–2000), once said, 'We are just 22 years old in terms of having a market economy – you cannot compare us to Hong Kong or Singapore' (Sanchata 2012). Empson and Webb write that 'This argument promotes the idea, not so much of "resource nationalism", but more of an image of an "innocent newcomer" that is learning the practice of contemporary international political economy: the importance of private contractual agreements and the detached yet supportive role of the State in underpinning that environment; the raising of private finance and financial governance requirements; and macroeconomic planning generally' (Empson and Webb 2014, 241). The learning experience of this 'innocent newcomer' from the contemporary international political economy and nationalistic responses to different consequences of the global economy has been a process of the indigenous shaping neoliberalism, not just how neoliberalism shapes the local (see also Tsing 2005) as many previous works depict.

Notes

1   At the subsequent trial, the court found that a state special security officer was responsible for the gunshot, not the protestors (see Chapter 5).

2 The square was originally named after the communist revolutionary leader Sükhbaatar Damdin, and the name changed to Chinggis from 2013 to 2016. The above incident occurred on the square when it was called Chinggis Square.

3 The peace was also interrupted on 1 July 2008, by protestors who resisted the results of the election. In the riot, police used tear gas and non-lethal weapons and killed five people. Unlike this riot, on 16 September 2013, it was the activists who brought arms in order to show their resistance against decisions of the central government and Parliament.

4 Uyanga is a woman in her forties who is a journalist and activist. She had been organising and leading dozens of nationalist movements and demonstrations against the state rulers, and was elected to the Parliament in 2012.

5 For the full video, see www.youtube.com/watch?v=eQiaMYSGdO0.

6 For other dilemmas in Mongolia, see also Badral Zulbayar (2015) and Mendee Jargalsaikhan (2018).

7 Many others also write about how other countries (Cordero et al. 2005, 1; Song and Woo 2008) experience similar economic and environment dilemmas in their own ways. International organisations, analysts and the media frequently address the economic slowdown triggered by environmental policy. Also, major surveys also suggest the same. For example, *Forbes Insights* survey shows that the United States regulatory environment has more impact on business than the economy (Moreno 2014).

8 Raúl R. Cordero, Pedro Roth and Luis Da Silva use a simple graphical model to show co-related growth of GDP and carbon dioxide emission. They argue that in order to meet both economic growth and environmental protection it is necessary to diminish the rate between pollutant emission and economic growth unit and the rate between resource consumption and economic growth unit (Cordero et al. 2005, 1).

9 By *State* they 'mean the Constitution and all the rules and actions of public services that flow from it' (Empson and Webb 2014, 232).

Introduction

This book presents the debates, discourses and reactions in Mongolia after the collapse of the socialist regime, which were triggered by critical economic situations, the development of the extractive industry, and environmental degradation that came about during efforts to craft the Mongolian nation-state (for crafting the state in Inner Asia see also Sneath 2006). It presents the emerging interactions and conflicts in what I refer to as the multipartite relationship of different agents. These agents and stakeholders in multipartite relationships use sets of concepts as discursive resources to understand, explain and justify their decisions and actions – by radically rethinking what it means to be a Mongolian, the importance of the nation, the proper form of state authority, the role of citizens, which forms of economy are just, and how nature, the environment, territory and sovereignty should be protected. The set of concepts that are mobilised include 'neoliberalism', 'capitalism', 'national economy', 'political independence', 'resource nationalism', 'alternative economy', 'state', 'environmentalism', 'tradition', 'pastoralism' and 'nomadism', and this book shows how they are contested in public culture and how they are assembled to form different ideological platforms with normative and therefore political implications. Accordingly, throughout the book, I treat these terms as concepts, perspectives and categories held and articulated by certain people towards certain entities for their own reasons.

The ethnographically driven multipartite scope that I employ in this book allows me to inclusively capture the breadth, and accurately portray the complexity, of Mongolia's political economy – including resources stretched across multiple local, national and transnational agents. Here, I use the aforementioned description to comment on existing interpretations of Mongolia's recent political economy and, further, to understand similar challenges in resource economies in different parts of the world.

Many works in the field of resources, mining and the state ask why resource-rich countries fail to reach prosperity. This has led to the development of the concept 'resource curse' to answer this question. 'Resource curse' refers to the irony that natural resource-abundant countries tend to have less economic growth, of which Mongolia can be an example. In other words, the term 'resource curse' is about a failure to benefit from resource endowment, experiencing Dutch Disease,[1] sovereign debt, bad governance, resource rents, myopic policy, corruption and mismanagement (see also Auty 1993; Sachs and Warner 2001; Gylfason et al. 1999; Krastev 2004; Ross 1999; Rosser 2006). Some acknowledge the same paradigm of resource curse to explain Mongolia's economic crisis. Indeed, to a certain extent, Mongolia reveals some symptoms of the so-called resource curse. For example, in 2011 Jeffrey Reeves elaborates in detail some of these symptoms in his article, 'Can Mongolia Avoid the Resource Curse?' In his elaboration, he identifies some key symptoms of the resource curse in Mongolia. Those include weak governance, non-transparency, lack of accountability, corruption and violation of mining legislation. He concludes that 'there is little reason to expect it will escape from the resource curse' (Reeves 2011, 182). I find that 'resource curse' is a generic way to understand different countries having similar experiences around the world. Gisa Weszkalnys (2011, 345) critiques the ability of the 'resource curse' to make sense of apprehensions of the past, present and future consequences of extractive industry developments. Following Weszkalnys, I seek to reveal the inner workings and components concealed in the 'black box of the resource curse' (Weszkalnys 2011, 356) by developing detailed illustrations of why and how Mongolian state rulers failed to bring prosperity.

Compared to a resource curse, 'resource nationalism' is a concept that is more appropriate to utilise to discuss Mongolia. Many scholars, such as Misheelt Ganbold and Saleem Ali (2017), use the latter concept to depict Mongolia. However, many others, such as Rebecca Empson and Tristan Webb (2014), Jargalsaikhan Sanchir (2016) and Julian Dierkes (2016), argue that 'resource nationalism' is a term that serves as a political tactic to place pressure on the nation-state of Mongolia. Similarly, John Childs (2016) and Natalie Koch and Tom Perreault (2018) complain about neoliberal bias and reductionism in the use of the term 'resource nationalism'. In Chapter 2, I argue that although it is not right to call it 'resource nationalism', it is also incorrect to argue that there is no nationalism in Mongolia's resource economy and policy. Instead, I suggest considering nationalist sentiments that appear in the resource business in the broader sphere of ideas of the nation in Mongolia.

Conflict is another theme that many people employ to explain resource economies. The conflicts in Mongolia's resource economy, however, are in many ways different from what Michael Watts (2004) presents in Nigeria or what Anthony Bebbington (2012a and 2012b) presents in Peru. Both of these authors address longstanding national, territorial, and historical and ethnic conflicts (Watts 2004, 71–2; Bebbington 2012b, 225). In Mongolia, conflicts between movements, companies and the state are not longstanding, historical or ethnic. Here, conflicts are not primarily between different ethnic groups or nationalities, or different class groups, or between minority and majority, or rural and urban, or central and peripheral populations. Instead, I argue that the conflicts emerge from the contradiction between nationalist and neoliberal ideologies. Since the collapse of socialism, different agents have promoted either nationalism or neoliberalism, but they could never completely eclipse the other (see Chapters 1, 2 and 5).

'Resource curse', 'resource nationalism' and conflict all attempt to understand and interpret the problems, challenges and conditions in the political economy of resources in Mongolia. However, this book frames the situation in a broader and comprehensive scope of nation building. Agents within the state, mining companies, donor organisations, and environmentalist and nationalist mobilisations often claim that their intentions are in the best interests of the country, nation and the environment. In other words, they all engage in different tasks of 'commoning', a process of creating and nurturing a community (or a nation in the case of Mongolia) that includes non-human agents (Blaser and de la Cadena 2017, 186; Linebaugh 2008, 279; Bollier and Helfrich 2012; Papadopoulos 2010). Yet the 'commoning' in Mongolia differs from the form that Peter Linebaugh coined. In the work of Linebaugh the 'common' refers to natural resources and the 'commoning' refers to 'relationships in society that are inseparable from relations to nature' (Linebaugh 2008: 279). Yet in this book 'commoning' is shown to concern not only the environment and resources but also the political and economic interests of the nation and the common good, such as matters of the sovereignty, political independence, national economy and the autonomy of the state. In that sense, the verb form 'commoning' refers to what Mario Blaser and Marisol de la Cadena call the 'invocation of the national common good' (2017, 185). In other words, in Mongolia, all acts of 'commoning' in the end serve the common agenda to build an independent nation and craft a sovereign state. For example, the

INTRODUCTION

rulers of the state claim that they supported the liberal economy and mining industry to save the national economy and to consolidate the political independence of the emerging nation-state as a matter of urgency (Chapter 1). The so-called 'resource nationalist' policymakers and technocrats claim that they supported the state control on natural resources to contribute to the security of the nation-state and to prevent foreign dominations (Chapter 2). Owners and operators of mining companies claim that they 'produce wealth' (*bayalag büteekh*) to contribute to the national economy and to fund the sovereign nation-state following the appeal of the liberal government (Chapter 3). Donors claim that they supported environmental movements to protect local democracy and civil society (Chapters 4 and 5). Activists claim that they intend to protect the common land, water and the environment abandoned by the state institutions (Chapters 5 and 6). All of these serve the sake of the public and national interests and common good to build an 'imagined community' (Anderson 1983).

Contest of Indigenisation and Neoliberalisation

In their study of social movements, mining and policy, Anthony Bebbington et al. (2008, 2902) argue that 'the institutions, structures, and discourses that govern asset distribution, security, and productivity are not pre-given. They are struggled over, re-worked, and co-produced through the actions and interactions of a range of market, state, and civil society actors.' The same argument helps to understand problems in Mongolia. To condense the complexity and breadth of the political economy of resources in Mongolia, I consider two processes that Bebbington et al. call the co-production – indigenisation and neoliberalisation – in the above-mentioned framework of nation-state. With the term *indigenisation*, I refer to nationalist and statist initiatives intended to transform elements to be ethnically, historically, traditionally and authentically Mongolian. With the term *neoliberalisation*, I describe the neoliberal policies that established the global capitalist free market economy in Mongolia. The central project to build the sovereign nation and the state neither fully acknowledges nor entirely rejects the two; instead, many agents in this book suggest the incorporation of the two by finding the right balance. For this reason, since the collapse of socialism, challenging procedures and experiences, along with political, economic, social and cultural debates have been required to find the delicate balance between indigenisation and neoliberalisation.

There are two related but different meanings of the word *indigenous*. As I mentioned above, one concerns the historical, traditional and authentic culture thought to represent a nation. The second is about minority peoples and communities of different races who are distinct from the politically dominant majority population. In the context of Mongolia, I adopt the former usage and reject the latter. The word *indigenous* is new to Mongolia, mainly because the population of Mongolia cannot be divided into indigenous and non-indigenous groups, unlike many other places that use the term *indigenous* (such as in America and Australia). My use of the word *indigenous*, therefore, does not intend to support the use of the term recently developed in Mongolia: that is, to introduce the United Nations' declarations on the rights of indigenous people, and to better allocate shares and benefits from mining operations to local people. I find such movements to be another form of indigenisation happening in Mongolia. The alternative form of indigenisation – a focus of this book – is about the making of the ethnic Mongolian people. Mongolians often use the word *mongolchilokh* or *mongoljuulakh* (to *mongolise*, or *to turn more Mongolian*) to talk about making things or beings more Mongolian. The concept permits and encourages people to convert anything foreign into something more Mongolian by enhancing features considered to be Mongolian, or using alternative local techniques to transform it.[2] In this sense, *mongolisation* is a process and a technique to make things happen or make things more indigenous, native and local. I argue that what happened to the political economy of resources in Mongolia was nationalist responses to transformations designed to indigenise or *mongolise* neoliberal policies and the capitalist market economy. As a result, neoliberal policies and the market economy become an unrecognisable hybrid for some people. For this reason, some observers may complain that Mongolia does not have free market and liberal economy, while some others argue that because consequences of neoliberalisation were not suitable for Mongolia, it needed *mongolisation* or what I call indigenisation.

With my use of the term *neoliberalisation*, I intend to respond to two related but distinct discussions of neoliberalism. The first considers neoliberalism as a 'conceptual trash heap' (Mair 2012) and is therefore 'an obstacle to the anthropological understanding of the twenty-first century' (Eriksen et al. 2015, 911). Taylor Boas and Jordan Gans-Morse 'document three potentially problematic aspects of neoliberalism's use: the term is often undefined; it is employed unevenly across ideological divides; and it is used to characterise an excessively broad variety of phenomena' (Boas and Gans-Morse 2009, 137). Following Boas and Gans-Morse, on 1 December 2012, in the Group for Debates in Anthropological Theory (GDAT), James Laidlaw and Jonathan Mair proposed a motion: 'The concept of neoliberalism has become an obstacle to the anthropological understanding of the twenty-first century' (Mair 2012). Later Mair wrote on his webpage that whenever the term *neoliberalism* is tagged, different usages 'seemed to be talking about quite different, even contradictory, things—the neoliberal tag seemed to add nothing' (2012). Mair continued, 'Or worse, it seemed portentous to invoke a whole global theory as a background explanation without doing the work of showing how "global forces" are linked to or expressed in the sort of "local" settings' (2012).[3] I acknowledge this argument, and I agree that cross-culturally, neoliberalism as a concept and a tag can be incoherent and even contradictory. Although I do agree with this claim, to write about the resource economy in Mongolia it is hard to abandon the use of the term *neoliberalism*. I also think that it is still possible to use it by making clear what it means in Mongolia. To overcome the burden of the term *neoliberalism* as something that can be empty and in flux, I shall clarify my use of the term, based on how Mongolians understand the current era. Mongolians often use the term *zakh zeeliin üye* (the era of the market; see also Sneath 2012) to talk about contemporary Mongolia, which is more about neoliberal policies and neoliberalisation processes that established the capitalist free market economy. The era of *zakh zeeliin üye* in many ways captures the framework of neoliberalism proposed by Manfred Steger and Ravi Roy (2010, 11). Following Steger and Roy, I take neoliberalism as three intertwined manifestations: an ideology; a mode of governance; and a policy package of deregulation, liberalisation and privatisation in the economy. In brief, in Mongolia neoliberalism as an ideology was employed by political rulers, and economic liberalisation policies they put into action established the capitalist market economy. I adopt the definition of Steger and Roy, not to test whether it fits in the case of Mongolia but to argue that neoliberal policies have been influential in Mongolia, at least in the range of the above manifestations.

The second understanding of neoliberalism considers it as an external and neocolonial power of international donor organisations, transnational corporations and capitalist states such as the United States (Harvey 2005; Harvey 2010, 28; Graeber 2011, 2; see also Peet 2003; Ferguson 2006; Ong 2006), and threatens the sovereignty of the nation-state (Sassen 1996). It is common in the literature of neoliberalism, nation-state, globalism and capitalism to consider neoliberalism and nation-state projects as contradictory. For instance, in the works that discuss neoliberal domination and neocolonialisation (Harvey 2010, 28; Graeber 2011, 2; see also Peet 2003; Ferguson 2006; Ong 2006; Harvey 2005), the death of the nation-state (Sassen 1996), or of 'resource nationalism' (Bremmer and Johnston 2009, 149; Vivoda 2009, 532; Maniruzzaman 2009, 81; Click and Weiner 2010, 784; Kretzschmar, Kirchner and Sharifzyanova 2010), neoliberalism for non-Western states stands as an external force that inflicts and threatens nation-states and indigenous nations (see also Bargh 2007). I do not argue against such approaches. My intent is neither to reject this claim nor to take this approach. I choose not to repeat this claim in the case of Mongolia; Morris Rossabi (2005) and Lhamsuren Munkh-Erdene (2012) already did some work to this end. In its place, I want to shed light on the employment of neoliberal ideologies in the process of nation-building and state-crafting in Mongolia, where indigeneity shapes neoliberal policies. Therefore, this book focuses not only on how global forces shape Mongolia but also on how the nation-building and state-crafting projects in Mongolia shape and make neoliberal policies incoherent or hybrid. The hybridity of neoliberal policies I present here is not the hybridity discourse about the neocolonial structures of power, embodied by ultra-liberalism and how international financial institutions (World Bank [WB] and International Monetary Fund [IMF]) appropriated and accommodated hybridity to achieve their own goals (Acheraïou 2011, 179). Instead it is about how indigenisation affects neoliberal policies and free markets.

In the framework of nation-building and state-crafting, for some political leaders neoliberalism became an ideological instrument employed in the national agenda and capitalist market economy became an opportunity to fund the nation and the state, as I mentioned before. In other words, some rulers of Mongolia considered marketisation, privatisation and deregulation to be not only neoliberal projects but also a mission that tied up with welfare of the nation to fund the emerging nation-state (see Chapter 1). Such claims make neoliberal ideologies nation-building policies of the nation-state. In this project, the nation-state rulers have some freedom to decide or debate on which conse-quences of marketisation to experiment with, which ones to attain and which ones to reject, although attempts to reject are always regarded as 'resource nationalist' (see also Joffé et al. 2009, 4; Domjan and Stone 2010, 38; Bremmer and Johnston 2009, 149; Wilson 2015, 399; Childs 2016, 539). In Mongolia, it can be regarded as indigenisation or *mongol-isation* by adjusting the alien supremacy, which does not have to fit in all of the principles to build the sovereign indigenous nation and the state. The *mongolisation* does not put neoliberal ideologies in a position of an alien power. For example, while some rulers of the state employed neoliberalism, and established the free market, others expressed critical approaches to neoliberal policies and attempted to control the process of marketisation; their intentions helped to indigenise the foreignness and unfitness of some consequences of neoliberalisation in Mongolia. Here, I mean the state control, regulation, navigation by limiting, adjusting and removing what is considered to be contrary, conflicting and threatening to what is called the *ünet zuil* (values), *yazguur erkh ashig* (original interests) and *bakharhal* (pride) of the nation. In other words, some Mongolian rulers, politicians, technocrats and activists attempted to fight against some neoliberal policies and control some consequences of marketisation (some were successful and others were not). In building the nation-state, these leaders often highlight different matters of the overall national project and use the notion of nation-building and state-crafting by 'commoning' to decide which consequences of neoliberal policies correspond or contradict other matters. Therefore, what shapes neoliberal policies and neoliberalisation processes is nationalist and statist political decisions that are related to the imagined nation-building (Anderson 1983) and state-crafting. In this vein, the issues of hybridity of neoliberalism – that are widely targeted by scholars – are related not only to the WB and IMF but also to varied agendas of different national actors.

Ascription of Ethnic and Civic Nationalisms

The understandings of the nation, nationalism and state are inseparable in Mongolia, and they have radically different meanings compared to Euro-American contexts (see also Gankhuyag 2007). In the historically constructed project to build a sovereign nation and its own state, the process of indigenisation on the one hand, and neoliberalisation on the other hand, significantly shape the concepts of people, nation, sovereignty, state and nationalism.

This book gives an account of how different agents in the state, mining companies and popular mobilisations endorse different discursive resources in processes related to political independence, sovereignty, environment and economy, and create routes to ratify nationalism. For example, some rulers of the state claim that they had to consolidate the independence of the state to end the domination of the Soviet Union, to prevent the domination of China, and to balance dependences by welcoming power of other countries, organisations and transnational corporations (see Chapter 1). Nationalist technocrats and politicians claim that they supported the control of the state to protect the sovereignty of the country from risks of private mining companies, which can cause damage to the environment and territory of the country (see Chapter 2). Environmental and nationalist activists consider that by protecting the environment, they perform the sovereign right and duty of the state (*tör*) to protect its territory in the absence of state protection (see Chapter 5). Lhamsuren's (2006) work points out that the underlining principles of all of these routes effectively endorse nationalism. That principle in Mongolia can be described as follows: 'Mongolian collective identity, ethnic, national, or whatsoever it might be labelled, made the Mongols see themselves as an inherent community or entity that had its own distinct origin, culture, lifestyle, homeland and ruling institution, and was thus destined to live as such' (Lhamsuren 2006, 92).

In this book, nationalism is largely used to describe nation-building and state-crafting. Although ethnic features are central to nationalism in Mongolia, it does not mean that the so-called civic nationalism is absent in nation-building. Socialist and postsocialist nationalism successfully incorporated the ideas of modernity such as urbanisation, industrialisation, democracy, human rights and globalisation. In Mongolia's nation-building, there are forms of ethnicised nation-protecting in addition to features of civic nation-building, although not in the sense of protecting the culture of minority groups or an ethnic other. Instead, it is about protecting the 'traditional' (*ulamjlalt*) 'cultural practices of past eras' (Verdery and Humphrey 2004, 17) of the Mongol nation. Or in the words of Lowell Barrington, 'in the case of ethnic-nations such [nation-building] policies would privilege the majority group at the expense of ethnic minorities' (Barrington 2006, 21).

In the absence of the majority and minority conflicts, the contradiction of the civic and ethnic features in contemporary Mongolian nationalism are the main points of contestation in the building of the nation and nationalism. The civic features endorse the free market, liberal economy, mining industry and democracy, while the ethnic features endorse the protection of the environment, 'tradition' and indigenous culture. In other words, for many Mongolians nationalism should embody both the civic and ethnic principles; it is impossible to completely abandon any of the two in the project to build a modern Mongolian nation (see also Brubaker 1999). The most challenging task is to find an accurate balance of the two – but it is possible for certain groups to support one more than the other. For example, environmental and nationalist movements tend to promote the ethnic (Bumochir 2018a, 106), while some Mongolian rulers, policymakers and technocrats promote civic tendencies, such as the neoliberal economy and mining industry. Using these features, some critique the former to be extremely nationalist, while others also accuse the latter of being not nationalistic enough. However, because of the inseparable fusion of the two kinds of nationalistic tendency, none of them can entirely ignore the other. In the following elaboration of the recent historical overview of nationalism in Mongolia, I will explain the historical and political consequences that produce a division between the so-called good and bad nationalistic tendencies, and how the boundary between civic and ethnic blurs.

It is obvious that the distinction between good and bad nationalism is a political matter. Barrington (2006, 10) writes that 'civic nationalism is often portrayed as good'. Unlike in Barrington's work, in Mongolia civic nationalism is not generally portrayed as good and ethnic nationalism as bad. In Mongolia, in addition to *ündesnii üzel* for nationalism, which means a view, perception and ideology of root and origin that can contain both civic and ethnic nationalistic tendencies, Mongolians also use another term, *ündserheg üzel*, which indicates an act to highlight, rank, dominate, discriminate and violate human rights. For example, in Mongolia those who promote neoliberalism often translate the term *resource nationalism* into *bayalgiin ündserkheg üzel* (see also Semuun 2013) not *ündesnii üzel*, to address its negative consequences. Apparently, such a division between good and bad nationalisms appeared after the collapse of the Soviet regime, at the time when Mongolia required a form of nationalism to ideologically support the nation-state on the one hand, while responding to the critique of bad nationalism on the other hand. Although this is how Mongolians commonly divide nationalism, materials in Chapter 2 show that sometimes it is impossible to identify whether an instance of nationalism is good *ündesnii üzel* or bad *ündserkheg üzel*. In other words, the good or bad division of nationalism is a political matter as I mentioned earlier and the same also applies to the Mongolian translation of 'resource nationalism'. Although it is translated as *ündserkheg üzel*, depicting some negative consequences on the national economy, for many it is also inseparable from *ündesnii üzel*, the so-called good nationalism.

The breadth and embedded nature of nationalism in Mongolia challenges what Barrington (2006, 4) calls the misuse of nation, the error made by equating it with 'state' or 'country'. Different terms can indicate different things, which of course is also true in contemporary Mongolia. However, as we have seen, not equating nationalism with state or country will not help to understand the nation (*ündesten* or *uls*), nationalism (*ündesnii üzel*) and the state (*tör*) in Mongolia. Rather, the embeddedness of nation and state in Mongolia arises from a particular history of a certain people and therefore the manner in which they generated nationalism.

The De-reification of the State

Antony Bebbington notes that in the political economy of natural resources the 'state is largely produced and moulded through social mobilisation, social conflict and the ways in which these processes are mediated' (Bebbington 2012b, 221). Along the same lines, in Mongolia conflicts between environmental movements and mining companies require state authorities, institutions and laws to turn to a mediator or a central agent. In this book, I provide an account of the ways in which different agents in the state institutions, mining companies and popular mobilisations produce and mould the state, and transform both the understanding and the institution of the state by what I call de-reifying the historical and cultural concept of the state.

The chapters of this book show how different agents in the mining companies, protest movements, NGOs and donors take the opportunity to engage with the state (the president, government and parliament, and institutions, decisions, law and regulations). The book presents narratives of key figures who have been central to political processes and debates. These different agents take one of two contradictory trajectories that I explained above: neoliberalism, free market and the liberal economy of resources; or nationalism, environmental protection and state control of natural resources. In other words, mining company representatives and liberal politicians maintain that the Mongolian state exceeded its power to rule and control the market, while environmental protestors and nationalist politicians complain about the absence of the state and argue that what is needed is to bring back state control in order to solve conflicts. For example, when I participated in the 'Discover Mongolia' annual investment forum of the MNMA in September 2015 – which appealed 'For Mining Without Populism' – many of the presentations complained that the Mongolian economy was not liberal, and that there was too much state participation and control. Jargalsaikhan Dugar, mining economist and the former president of the MNMA, presented a paper titled 'Business Freedom in the Mining Industry'. He noted that the state violation of the business freedom in the mining industry abuses some of the basic principles of Mongolia's constitution; that is, to provide human rights and private property rights (Jargalsaikhan 2015). However, I heard the opposite many times – for instance, from some environmental and nationalist activists and protestors. For example, in Chapter 6, another prominent figure of my book, Munkhbayar Tsetsegee, the founder and leader of the river movements, winner of the Goldman Environmental Prize in 2007[4] (known as the 'Green Nobel Prize') for his achievements in closing down gold mines, and the National Geographic Emerging Explorer in 2008 for successfully protecting rivers in Mongolia, develops an argument on the absence (rather than excess) of state protection of the environment, territory and people (see Chapter 5).

Epifanio San Juan describes how after World War II, when speaking of the nation-state, it 'became axiomatic for postmodernist thinkers to condemn the nation and its corollary terms, "nationalism" and "nation-state", as the classic evils of modern industrial society. The nation-state, its reality if not its concept, has become a kind of malignant paradox if not a sinister conundrum' (San Juan 2002, 11). This analysis of the nation-state and nationalism can be a surprise for many Mongolians whose understanding of the nation-state is completely different, as we have just seen. In Mongolia, *nation-state* mostly refers to what Barrington (2006, 21) calls the 'ethnic-nation-state': 'The ideas of the nation as an ethnic nation and the state as a nation-state combine to produce an "ethnic-nation-state"'. He also writes that in the minds of nationalists, the state, as a nation-state, exists for the benefit of the nation. As such, if the nation's cultural identity is threatened, state policy must be adopted to protect the culture from the threatening 'other'. Apparently, although neoliberal and nationalist agents demand opposed things from the state, they both share state protection as a focal point. For example, for liberal rulers who launched capitalism and an international economy in postsocialist Mongolia, the state had to protect the national economy to fund the nation-state (see Chapter 1). For environmental protestors, it was more important for the state to protect its environment from mining-induced damage, as it is an important part of the state's duty to protect its territory. Although liberal and nationalist agents support contradictory trajectories, they all agree on the importance of the fact that the Mongolian state is an ethnic-nation-state that is expected to protect its people, culture, territory, economy and sovereignty. In brief, this agreement also demands that the state follow three principles: a political independence that is free from any external powers and influences;

visibility and presence of the state that declares its sovereign right; and the homogenous Mongol nation entitled to claim the rule of the state. The widespread cultural acceptance of the state power and its protection helps to explain why and how statism and discursive resources related to the nation-state and its security often appear as the most powerful ones to shape neoliberal policies and the market economy. In that sense, the de-reification of the state provides different agents not just an opportunity to participate and influence the state, but also to maintain the state by demanding its protection on the national economy and the environment.

Mining and Popular Mobilisations

There are many works in the field of mining, resource, environment, activism, governance, state and capitalism that present conflicts between mining, protest movements and the state from different parts of the world,[5] such as Indonesia (Tsing 2005[6]; Welker 2014[7]), Colombian Pacific (Escobar 2008[8]), Nigeria (Watts 2004[9]), Bangladesh (Gardner 2012[10]), Argentina (Shever 2012[11]), Papua New Guinea (Kirsch 2014[12]; Golub 2014[13]), Peru (Li 2015[14]) and Bolivia (Andreucci and Radhuber 2017[15]). Much of this literature has three striking similarities. First, they often utilise a dual or triad 'stakeholder', 'agency' and 'actor' framework referring to a corporation, community and the state (Ballard and Banks 2003, 290; Watts 2004, 54; Richardson and Weszkalnys 2014, 9–11; Welker 2014, 4; Gilberthorpe and Rajak 2016, 189). Second, these works depict local communities as powerless and corporations as powerful. Third, they often find that the state or government generally supports the mining industry. This book argues that these three commonalities are not present in Mongolia.

Regarding the first point, Emma Gilberthorpe and Dinah Rajak (2016, 187) suggest that 'at the heart of debates about the resource curse … lie persistent questions about the relationship between extractive TNCs [transnational corporations], governments of resource-rich countries, and local populations (or stakeholders)'. However, this study of Mongolia shows that there are more than two or three groups and parties involved in the resource conflict, which I refer to as a multipartite relationship. I argue that the situation often involves many disparate groups and people with complicated relationships to each other; in addition, those different parties, people and their divisions are not stable. For example, in the so-called local community, there are the local government employers, local residents who live in settlements of the administrative unit centre, and local herders who live in the countryside. In the protest movements, there are varied groups from different places, sometimes with distinct purposes or methods of protest, all of which sometimes unite and split. Different foreign donor organisations from Germany, Japan and the United States have supported popular mobilisations, but some have had severe conflicts with the river movements. State stakeholders also have significant distinctions, such as the central government, parliament, the president of Mongolia, political parties and other politicians. Mining companies also cannot be considered as one group of agents, as there is a vast difference between small- and medium-scale mining companies and transnational corporations. One should not lump all of these groups and individual agents into two or three groups. This book provides proper consideration of them as distinct parties, and agencies with different interests, visions, principles and purposes; and through it all, their relationships to each other remain fluid.

Second, the literature mentioned above usually portrays how resource extraction causes environmental problems for the powerless, frontier, and minority, tribal, indigenous and local populations (Tsing 2005, 3; see also Chapter 4), while often depicting powerful transnational corporations foreign to the local population and the nation-state (Ballard and Banks 2003, 293–4; Kirsch 2014, 1; see also Chapter 3). In Mongolia, the local population that faces environmental problems is not an ethnic minority, tribal community, or indigenous group of a different race. The context of the nation in Mongolia does not put the local population in the position of the peripheral, powerless and minority other. Moreover, in many cases of environmental nationalist protests in Mongolia, local governments and authorities started and promoted strikes and movements against those promoting, advocating and operating mines. Movement leaders often run for parliamentary elections, and sometimes become parliament members. They lobby politicians and environmental laws to stop mining damage (see Chapters 4 and 5). Therefore, the conventional presentation of the powerless and peripheral whose voices are not often heard in national and global arenas is not the basis of my work. The power of protestors places mining companies in a position other than the conventional position of the powerful. Also, the image of 'powerful' does not apply to all corporations, especially to the small- and medium-scale mining companies, investors, and Mongolian corporations, which are also the focus of this book. Most of the literature rarely discusses the situation of small- and medium-scale corporations, and national mining companies that are powerless in the face of the powerful local resistance movements, state policies and regulations to protect the environment and local residents. My work attempts to deconstruct this generalised image of mining corporations and unmask the complexity of the issue. I argue that the overall transnationalisation of corporations tends to obscure the challenges and struggles of many other small- and medium-scale corporations and national mining companies.

Third, despite the power of communities and corporations, state officials often intervene in the conflict and make political decisions sometimes in support of national/local movements and sometimes in support of mining companies, which makes the powerless and powerful designations further irrelevant (Bebbington et al. 2008, 2891). Unlike in most countries where the central government or the state is committed to expanding the resource industry and the mining economy (Coronil 1997; Bebbington 2012a, 10; Andreucci and Radhuber 2017, 280),[16] the nation-state rulers, the central government and the Parliament of Mongolia do not always sustainably support the mining industry and the neoliberal economy, as most of the works mentioned above delineate.

Structure of the Book

The chapters of this book follow a chronological order of events that happened to and in Mongolia since 1990. Throughout, I provide the accounts and narrations of historical and contemporary events by key figures involved in varied parts of Mongolian society and political debate. These accounts are an attempt to provide a form of counter-narrative from the Mongolian perspective to decentre the dominance of the 'resource curse' perspectives outlined above. Three chapters in the first half of the book are about the emergence of neoliberalism, liberal economy, nationalist debates to control the resource economy, and the expansion and challenges of mining companies. The three chapters in the second half of the book are dedicated to the environmental and nationalist popular mobilisations, their achievements, challenges and ideas about the state, environment, pastoralism and national identity.

Chapter 1 details the political economic situation of Mongolia after the collapse of socialism, and the solutions of the state (parliament, central government and the president) to help the situation. One important solution was the opening and development of the mining sector, led by the first president of Mongolia and the democratic union. This solution has had wide-ranging impact and consequences, including the relentless growth of mining capitalism. The five chapters that follow present different consequences, responses and resistances to the different impacts of mining, capitalism and neoliberal policies that were established in Mongolia. Chapter 1 focuses on the reification of the concept of the national economy and the anxiety of the loss of political independence as narrated by some Mongolian rulers, such as Prime Minister Byambasuren Dash and President Ochirbat Punsalmaa, who ruled Mongolia after the collapse of socialism in the 1990s. They discuss the problems and challenges related to political independence and reification of the national economy to justify the establishment of mining capitalism in postsocialist Mongolia. According to their justifications, they were forced to employ neoliberal ideologies and establish capitalist market economy as an urgent solution to economic instability. I argue that political leaders marked the political independence of Mongolia in order to reify the national economy and this marking contributed to the nationalist building of the nation and crafting of the state.

Chapter 2 presents how some nationalist politicians and technocrats resisted the liberal mining economy supported and established by those introduced in Chapter 1. This chapter introduces Khurts Choijin, a former geologist and mining minister in the times of socialism, who pioneered movements to support the state control of natural resources that were intended to constrain the operation and business of mining companies and the liberal principles of the free market economy. Literature on resources and mining commonly depict Khurts's position and other similar approaches as 'resource nationalist'. Instead of rushing to 'pejoratively label' such positions as 'resource nationalist' and 'populist' (Myadar and Jackson 2018: 1), this chapter attempts to understand the cultural, historical and political influences in such positions that are often ignored in the 'resource nationalism' literature. In response to such critiques, Khurts's narrations deploy an indigenous concept of an alternative economy and articulates historical narratives of state control and protection. Taking cultural and historical discursive resources into consideration, this chapter explains that beyond the so-called 'resource nationalist' positions there are attempts to balance different schools of thought and policy. Also, although this chapter declines to employ the term 'resource nationalism' to discuss Khurts, it argues that Khurts's position was nationalist in the broader sense of nationalism in Mongolia. As such, the chapter underlines the wide-ranging scope and distinctiveness of nationalism in Mongolia, in contrast to the reductionist tag *resource nationalism*.

Chapter 3 introduces three small- and medium-scale gold mining companies that started around the 1990s and contributed to the national economy. This chapter narrates how they navigated statist and nationalist initiatives and policies, along with the critiques and resistances made by popular mobilisations. Out of the three described, the first stopped its mining operations, enlarged its other businesses in the field of agriculture and food production, and started other businesses. These manoeuvres helped the company to transform its corporate identity into a national food producer. Although the second one used similar navigations, unlike the first one, the company made a successful entry into national politics by establishing a political party, winning an election and establishing a coalition government with the help of the Democratic Party. However, the joint forces of political opposition and resistance movements against the company and its political party, and pressures of statist and nationalist initiatives forced the company to abandon its mining business. The owners of the third company in this chapter navigated and continued mining by subcontracting and making their businesses non-transparent using family and other networks. These companies represent those that were later known as *bayalag büteegchid* (wealth producers) who contribute to nation-building and state-crafting by assisting the national economy. As wealth producers – a title created in response to complaints and resistances of the destruction and exploitation of mining companies – they also reify the national economy and build the nation-state in their own nationalistic ways, much as political rulers did as shown in Chapter 1.

Chapter 4 presents popular mobilisations initiated by local government employees due to lack of water, and as peaceful local river movements against the destruction by mining operations in 2000. Those mobilisations started as an advocacy group of local and national elites and international donors that was later transformed by its leaders as local herders' grassroots activism. It produced different discursive resources; promoted the right to resist mining companies and neoliberal policies; and established recognisable ways to protest, with which local people might be able to stop mining operations or extract some wealth from mining companies or the state. This chapter also explores how the movement successfully achieved national and international recognition with the support of some donor organisations. However, the alliance between the protest movements and donor organisations ultimately collapsed due to disparate interests and agendas. All of the above achievements of popular mobilisations show, as I mentioned before, that popular mobilisations in Mongolia do not present local, peripheral, tribal, indigenous or minority voices of the powerless communities but powerful national and international forces that fight against mining destructions and the state, as the next chapter illustrates.

Chapter 5 concerns the manner in which the friction of local activists and international donors changed the discursive resources, emphases and rhetoric of the popular mobilisations. The emphases and rhetoric in the mobilisations expanded from mining to the failures of the state institutions, policies and rulers to protect its people, environment and territory. The use of weapons and explosives in the protest movements and their nationalistic discursive resources turned some of the river movements into the internationally known violent nationalist forces many understand them to be: locked in a seemingly endless battle against the state, international donors and mining companies. Of course, another way to look at the river movements is as they see themselves: as desperate and loyal citizens who sacrifice their lives for the protection of the environment or as an expression of the sovereign right of the people of Mongolia in absence of state protection. Those nationalist popular mobilisations were successful in closing down mines, lobbying for different laws and political decisions to protect the environment, and winning trials against mining companies and the government. However, their final, desperate attempt to seek attention by bringing rifles, grenades and explosives to a staged demonstration led to the imprisonment of five river movement leaders for committing terrorism, possessing arms and extorting mining companies. This chapter argues that activists de-deified the state to free state regulations, institutions and officials from the culturally salient legacy of the deified power of the state (*töriin süld*) to critique neoliberal policies, resist political decisions, blame state institutions, and to bring the state back by lobbying and amending laws, protecting the wellbeing of the environment and people and by radically rethinking the state.

Chapter 6 describes activists' responses to the accusation that these movements do not exist for the sake of the environment but are instead made up of green terrorists who extort mining companies. To respond to this accusation and justify popular mobilisations that he led, Munkhbayar employs the 'traditional' nomadic herder and pastoralist identity, way of life and cosmology. This chapter follows the actions of Munkhbayar, as he rethinks nomadism, pastoralism, national identity, environmentalism and state versus neoliberal policies and capitalist markets in Mongolia. He reflects on many important questions, such as what it is to be a human being or a Mongolian, what it is to be a nation, and what it is to have a sovereign state. To answer these questions, he imaginatively fashions what I call an *original environmentalist society* of the Mongol herders who follow the law of nature (*baigaliin khuuli*). Moreover, this understanding of the original environmentalist society helps popular mobilisations to respond to a different accusation – for example, the suggestion that mobile pastoralism is unsustainable and herders degrade the environment more than the mining. The image of pastoralism as an ancient and environmentally harmonious heritage helps 'herder activists' to reject such accusations.

Notes

1   'Dutch disease' is a term in economics that indicates an economic dependency on a certain sector such as mining. The term was first coined by *The Economist* to describe the economy of the Netherlands in 1977 (*Economist* 1977).
2   For example, Mongolians often complain that foreign laws and conventions adopted by Mongolia need some *mongolisation* – that is, the law must be changed to make it more fitting to conditions in Mongolia.
3   Also, many anthropologists find similar problems in their study of neoliberalism. For example, Laura Bear (2015, 7) writes that 'Anthropologists have shown that neoliberalism is not a single coherent project, but an assemblage of techniques and institutional structures' (see also Collier 2009; Ong 2006; Murray-Li 2007; Shore and Wright 1997; Shore, Wright and Pero 2011).
4   The Goldman Prize is a prestigious environmental prize. In 1989, Richard N. Goldman (1920– 2010) and his wife, Rhoda H. Goldman (1924–96), established the Goldman Environmental Prize. The prize honours the achievements and leadership of grassroots environmental activists worldwide.
5   Richard Howitt, John Connell and Philip Hirsch (1996) discuss the binary or triad relationship between a corporation, community and state in the case of Australasia, Melanesia and Southeast Asia.
6   Anna Tsing (2005) writes about the collaboration of Japanese tree trading companies and Indonesian politicians that resulted in forest destruction and the resistance of local indigenous Kalimantans, and discloses congeries and friction of the local and global.
7   Marina Welker (2014) seeks an answer to important questions: What is a corporation? What does it do to the local population? She explores audit, social responsibility, and state rulers' relationships with the American copper and gold mine corporation in Indonesia.
8   Arturo Escobar (2008) discusses capitalist development attempts to appropriate the rainforest and extract resources, and indigenous and environmental movements and their networks in the Colombian Pacific.
9   Michael Watts (2004) talks about armed movements, violence, indigeneity, nationalism, governmentality, multinational oil corporations and the struggle of the federal state in Nigeria.
10   Katy Gardner (2012) writes about multinational mining company gas plants and discordant narratives of dispossessed landowners, urban activists and mining officials in Bangladesh.
11   Elana Shever (2012) examines protests against the world's two largest oil companies, Shell and Exxon, and conversion from a state-controlled to a private oil market in Argentina.
12   Stuart Kirsch (2014) focuses on the gold and copper mine environmental disaster and conflict between corporations and indigenous peoples, advocacy groups, and lawyers in Papua New Guinea.
13   Alex Golub (2014) writes about the relationship between an indigenous group and a large international gold mining company in Papua New Guinea.
14   Fabiana Li (2015) writes about corporate mining, activism and state advocacy of the extractive industry in Peru and analyses the definition of mining pollution, and the concept of equivalence and compensation.

15 Diego Andreucci and Isabella M. Radhuber (2017) present the anti-neoliberal strike against the largest multinational oil corporations in Bolivia and the state's legislation and promotion in order to expand mining and protect the environment at the same time.

16 This general depiction is not always the case, for example in work mentioned above of Diego Andreucci and Isabella M. Radhuber in Bolivia, where many of the advanced law and regulations to protect the environment and local population passed. However, in a different article Andreucci (2017) argues that central government's acceptance of the indigenous and environmental concerns was a 'passive revolution', in the words of Antonio Gramsci (2000, 261) which weakens or paralyses opponents by taking over their concerns and leaders.

1

The Reification of the National Economy

This chapter presents the historical narrations made by some of the political rulers who were involved in the transition to capitalism in Mongolia in the 1990s. Their narrations often address the problems, difficulties and challenges related to the political independence and national economy of the country. They tend to reveal that politicians in the 1990s considered the political independence and national economy of Mongolia to have been in an extremely precarious state. Given the influence and domination of the two neighbouring powers, Russia and China, these political leaders as well as many other Mongolians commonly view Mongolia's political independence to be something difficult and challenging to achieve and preserve. I consider such concerns as a form of precarity identified and possibly imagined by political rulers. Second, regarding the national economy, they make an analogy between the political independence and the national economy, and in much the same way they depict different difficulties and challenges in the national economy of the emerging nation-state of Mongolia. This is another form of precarity identified and imagined by those political rulers. According to them, after the collapse of the Soviet Union and the economic system in the Soviet bloc, Mongolia was left in a critical economic situation to fund the nation-state emerging from the ruins of socialism. In such a critical situation, political leaders presented the national economy to the public as an entity that should be prioritised to fund the nation-state to support its vulnerable political independence, which I refer to as the reification of the national economy. To find a solution to secure the independence and assist the national economy, those political leaders welcomed the free market and democracy and the so-called 'third neighbours' including the US, Japan and other powers in Western Europe. This chapter intends not only to give a historical account of that period but also to provide a counter-narrative to the account that Mongolia was a passive victim of neoliberal forces of America and its allies (see also Bumochir 2018b). In this line, some consider the presence and participation of these third neighbours in Mongolia's reform not a 'contribution' but a 'domination and influence' that makes Mongolia dependant on them (Rossabi 2005; Munkh-Erdene 2012, 65). However, the presence of third neighbours and international donors that were invited and welcomed by some political leaders suggests something different from what some narratives depict – for instance, the one developed by Sara Jackson (2015) on how transnational corporations and external powers build the nation-state in Mongolia. These depictions present one side of the story and miss the deliberate intentions of the individual Mongolian political rulers' agendas and contributions in the process of building the nation-state, which this chapter demonstrates.

This chapter also shows how those political rulers generate nationalism from narratives of the precarious political independence and the reification of the national economy. In the first chapter of an edited book on nationalism after independence in postcolonial and postcommunist states, Lowell Barrington focuses on the question of how political elites 'maintain the nationalist movement after its ultimate goal – independence – has been achieved' (Barrington 2006, 14). He writes, 'While nationalist elites will be exhausted, ecstatic, or just pleasantly surprised when independence comes, they will also generally seek to continue the momentum of the nationalist movement. As not only something that they believe in but also their ticket to power, nationalists will search for ways to keep nationalism alive. But since the nationalism can no longer be about achieving independence, it must be transformed' (Barrington 2006, 14–15). Subsequently, he introduces two possible transformations of nationalism, both of which relate to the issue of territory control. The first is the 'homeland claims of minority groups in the state' and the second are the 'claims on its homeland by other nations outside the state' (Barrington 2006, 16). He notes that 'few nations in the world in control of their own state do not face one of these homeland problems' (16). Postsocialist Mongolia is a recent nation-state that falls into those few nation-states of Barrington that 'do not face one of these homeland problems' and national elites manage to develop and trigger nationalism in a somewhat divergent manner. Since the 1990s, Mongolian political discourses have tended to further consider the state of *tusgaar togtnol* (political independence), which literally means 'a separate or autonomous existence', to be precarious not only in regard to the issue of territory but rather in reference to the 'national economy'. This chapter shows that in Mongolia, anxieties regarding economic independence – or fear, worry and nervousness of the loss of the independent status of the nation and the state – serve to interfere with the very goal of political independence. Instead, the focus shifts to the national economy and economic independence, which becomes the terrain of political struggle for the movements working for national autonomy. As will be shown below in interviews with political leaders and their historical narrations, national economy is the pillar of national independence, which funds the state and prevents economic and financial dependency upon foreign countries and institutions. In this way, discourses concerning the national economy recreate a form of nationalism.[1] Political discourses of independence and the national economy that emerged in the process of building the modern nation generates *ündesnii üzel*, a Mongolian term for nationalism that means root- or origin-centred ideology. This process of nation building, and the sentiment it generates, in turn supports the liberalised mining economy and prioritises issues of the national economy.

There is a great deal of work exploring how concerns of political independence and sovereignty generate nationalistic sentiments and a substantial amount exploring how these processes might reify the national economy. However, there is a remarkable absence of material examining how the reification of the national economy might itself lead to particular forms of nationalism, an absence this chapter seeks to counter. Hannah Appel (2017, 294) begins an article on the ethnography of the national economy with the IMF list of the 'World's Best Economies'. Winners in 2013 included South Sudan – the world's fastest growing economy – and Equatorial Guinea – the economy with the most investment. This was in contrast to the previous year (i.e., 2012) which included Libya – the world's fastest growing economy – and Mongolia – the economy with the most investment (Riley 2012). She then asks, if these are the nations of the world with the 'best' economies, what then *is* a national economy? Appel argues that national economy is an imagined 'object of the future and a justification for the constant deferral of the present' (Appel 2017, 294), which political rulers, donor experts, and economists use to justify their political decisions and actions in Equatorial Guinea. She also presents a broader history of the national economy in Europe as an epistemological project of the state, born in a geopolitical moment of Western independent nation-states to manage the Great Depression, mark their sovereignty, and respond to shifting global orders (Appel 2017, 297–9; see also Stiglitz, Sen, and Fitoussi 2010; Vanoli 2005). The below historical narrations show how the epistemological project of the state in postsocialist Mongolia (national government, parliament, prime minister, and president) targeted at the national economy demarcates the sovereignty of Mongolia. However, unlike in Appel's work, it also shows how this marking reinforces the making of the state and nationalism (in the sense of *ündesnii üzel*). The sheer volume and variety of difficulties and urgencies in postsocialist Mongolia became the main ground to identify precarity and shape discourses of independence, sovereignty and national economy in the agenda to build the nation-state.

This chapter has two sections. The first focuses on political discourses that identify precarity in the political independence of Mongolia in the immediate aftermath of the collapse of socialism. This section revisits the political discourses of independence and the third neighbour policy and highlights its importance in the justification of the political decisions to ally to the United States and other so-called third neighbours. The second section is about identification of precarity in the national economy to fund the emerging nation-state, where natural resources play a pivotal role in addressing perceived precarity in the national economy. As a result, what Stuart Kirsch (2014) calls 'mining capitalism' was successfully established in Mongolia (see also Plueckhahn and Bumochir 2018). There was an urgent drive to assist the national economy in order to strengthen the political independence and sovereignty of the emerging nation-state by funding it, and this became the main logic to justify why the state (government, parliament, and the president) was obligated to promote neoliberal transformations in the 1990s (Bebbington 2012a, 10). I do not seek to question the historical accuracy of the claims of emergency in terms of independence and the national economy, but rather to explore how the identification of precarity in these issues constructs a political discourse that politicians use to justify the establishment of 'mining capitalism'.

Political Independence

Russia and China, two political and economic superpowers, sandwich Mongolia geographically. There is much historical evidence that demonstrates how and to what degree Russia and China (and some third powers, such as the United States and the United Kingdom) have influenced the political independence of Mongolia.[2] In the period following the end of socialism, to balance the influence of China and Russia, Mongolian politicians developed the idea of 'third neighbours' by inviting other powers to protect Mongolia from the political domination of Russia and China. The invitation of the third neighbours has been done by considering the independence of Mongolia to be weak or at risk of loss and therefore needing constant protection. Depictions of potential threats in the narrative of independence, which has been a by-product of intentions to glorify those who fought for independence, creates a fear of the loss of independence and warns Mongolians to prioritise political independence. Subsequently, such fears generate a form of anxiety of independence among Mongolians. As I find in this chapter, the circulation of anxiety in the population is a justification to politically mobilise by identifying precarity in the independence and reifying the national economy. For example, this anxiety is a key factor for some political leaders to justify why Mongolian rulers had to embrace America and capitalism (see also Bumochir 2018b). They also consider the precarity of independence and the importance of third neighbours as something that is real and actual.

Yet, not all Mongolian people agree that this policy is inevitable. Munkh-Erdene Lhamsuren, a Mongolian anthropologist working at the Max Planck Institute in Germany, argues that the precarity of Mongolia's political independence between Russia and China in making the third neighbour policy is an 'ideological construct' (*üzel surtal*).[3] According to Munkh-Erdene, Mongolia does not need third neighbours to establish political independence. He believes that even in a worst-case scenario, neither of the two superpowers would let the other take over Mongolia. In this sense, it is the interest of the two states and their balance of power that secures Mongolia's political independence. Therefore, for Munkh-Erdene, the anxiety of independence and the creation of third neighbours is a politically constructed ideology to justify the establishment of a particular form of capitalist markets and increases the influence of states, such as the United States. Only a few days after I spoke with Munkh-Erdene, I coincidentally watched a short video in which the Mongolian political scientist D. Bolor made a similar claim. The video was posted in a closed Facebook group named 'Xiongnu-Xianbei, Mongol-Gokturk', which are names of the historical empires and peoples associated with the region of Inner Asia. The profile picture depicts portraits of some historical emperors and includes accompanying text, which says, 'MONGOL is GRAND NATION. Mongols are originated from "Tengri" heaven'. In the video, Bolor can be heard saying, 'The so-called revolution in 1990 was not a revolution, but it was a *coup d'état* organised by the American and British intelligence . . . Since they control media and banking system, and they influence the economy and politics . . . To hide the fact that political leaders made Mongolia their colony, they created a virtual neighbour called the third neighbour; a concept that does not actually exist. Because a neighbour is a country that shares a physical border, but it can never be other countries that reside overseas on the other side of the world' (Tömör Temür Tamerlan 2018).

While the inevitability of the third neighbour policy may be up for debate, what the above views and further materials I present in the chapter exemplifies is that political leaders tend to actively inflame anxieties regarding Mongolia's political independence. Narration of those difficulties helps them to justify the established mining capitalism and political-economic alliance with America and Europe. Regardless of the facts,[4] the discourse of the political independence can be a historical and political construction that is available to use as the explanation, understanding and justification for different political decisions. The research materials that follow show how different individual political leaders of Mongolia interpret difficulties of political independence and present the situation as precarious.

Byambasuren Dash, the last prime minister of the MPR, had direct experience with political matters related to independence in the early 1990s. Now retired, this former prime minister remains quite popular and often appears on television and other media to comment on the national economy and politics. At the same time, many people also blame him for establishing the free market in Mongolia by supporting privatisation, price liberalisation and so on. Byambasuren's narrative is an interesting and important side of the story of Mongolia's transition: he was an economist with experience of high political office during socialism – including as deputy head of the Council of Ministers (1989) and prime minister (1990–92) – and was also the person who led the country in times of crucial change.

Due to the desperate situation of the country, in October 1990, only a month after he became prime minister, Byambasuren sent requests to the World Bank and the IMF to initiate membership. Caroline Humphrey (2002, xvii) describes the similar situation in the Soviet Bloc as follows: established institutions were disintegrating and decaying, which urged decisive and immediate reactions and the establishment of a new political economy. In reference to Mongolia, David Sneath (2002, 194) writes, 'Soviet support amounted as much as a third of Mongolia's GDP or more, and it was reduced in 1989 and stopped altogether in 1991'. Following Sneath, Gantulga Munkherdene (2018, 375) explains 'Unemployment and poverty spread, thrusting the economy into crisis. Due to this crisis, in 1991 Mongolia became a member of the International Monetary

Fund, the World Bank, and the Asian Development Bank, and started to welcome their assistance and recommendations'. In an attempt to find solutions other than sending requests to donors, Prime Minister Byambasuren Dash visited Washington, DC, in May 1990, just two months after the Politburo (*uls töriin tovchoo*) had stepped down and the establishment of the new government (Addleton 2013, 38; see also Buyandelgeriyn 2007, 129). Uradyn Bulag explains such visits in this way: 'Mongols rejected the Soviet Union which was the traditional guarantor of Mongolia's independence and had to face China alone' (2017, 121). Moreover, the 'Mongols were paranoid that an impoverished Russia might sell Mongolia to China' (2017, 212). When I met Byambasuren on 12 May 2016 he told me that Russia and China agreed in November 1990 to 'not use Mongolia against each other and not to let any third countries use Mongolia'. He also explained that in a critical situation such as this, it was dangerous to continue socialism because this would have brought Mongolia closer to China. At the same time, China had a vested interest in 'embracing' (*tevrekh*) Mongolia, as evidenced in many different sources that demonstrate Chinese claims that Mongolia should be a part of China (see also Bruun and Odgaard 1996, 39; Sanders 1996, 222–3). For Byambasuren, the anxiety was felt when he met Li Peng, Premier of the People's Republic of China (PRC), in Beijing on 19 June 1991. Byambasuren said that Li Peng repeatedly mentioned to him that 'China is building socialism'; according to Byambasuren, the implication was that Mongolia should join China. Therefore, he felt that it was urgent to replace the collapsing guarantor – the Soviet Union – with the United States.[5]

According to Byambasuren, while China revealed its aim to influence the MPR, the rulers of the Soviet Union did not simply allow Mongolian rulers to replace the Soviet Union with the United States. This is counter to the widespread understanding that the rulers of the Soviet Union freed Mongolia from its influence without hesitation. However, Byambasuren explains the situation based on information and materials that were never publicly revealed. He told me that in December 1990, the Politburo of the Soviet Union passed a top-secret decision to politically and economically sanction Mongolia, a reaction to the MPR's decision to abandon socialism and the Soviet Union. On the following day, the ambassador of the Soviet Union to the MPR visited the office of G. Ochirbat, chairman of the Mongolian People's Revolutionary Party (MPRP). Here, he read aloud the decision of the Soviet Union Politburo and left without leaving a copy of the document. Later, Byambasuren heard of this sanction from Ochirbat and realised that the sanctions were already in effect when the Soviet Union delayed its supply of oil to Mongolia in August 1990. The consumption of petrol in the MPR was then (and still is) entirely dependent on the Soviet Union (now Russia), which made this sanction highly effective. According to Byambasuren, in August 1990, the Soviet Union was supposed to supply 34,000 tonnes of petrol to the MPR, but instead Mongolia received only 17,000 tonnes. Following this, the Soviet Union completely halted its petrol delivery. Byambasuren described how the MPR was left in a vulnerable situation and almost completely depleted its national petrol reserves. To help alleviate the issue, when Byambasuren visited the United States a second time in June 1991, he informed James Baker, then-US Secretary of State, about the critical condition of Mongolia and the sanctions of the Soviet Union. Baker considered the situation seriously and told Byambasuren that he would warn Mikhail Gorbachev, the last president of the Soviet Union, that US support to the Soviet Union could not be secured if the Soviet Union continued its sanctions against the MPR. Byambasuren understands that it was only as a result of Baker's talk to Gorbachev that, on 27 March 1992, Mongolia finally received oil from Russia.

The sanction – if it existed as Byambasuren talks about it – was issued just a year before the dissolution of the Soviet Union. Further, Mongolian rulers' attempts to diversify its dependence and crystallise its guarantors started in the mid-1980s. MPR rulers began to reveal failures in its economy and launched a search for alternatives (Bulag 1998, 17; Rossabi 2005, 8); officials of the MPR also made approaches to the United States during the late 1980s (Addleton 2013, 39). Consequently, the Soviet military withdrew from the MPR in 1987. Byambasuren mentions that the sanction was probably a reaction of the rulers of the Soviet Union to those attempts revealed by the rulers of the MPR.

Not long after Byambasuren's first visit – on 4 May 1990 – Deputy Assistant Secretary of State Desaix Anderson affirmed that the United States had been 'presented with unique opportunities to be supportive of positive developments at a turning point in Mongolia's history and at a time when their leaders are looking to us for assistance' (Addleton 2013, 38). It was an important point for the United States to highlight that Mongolian leaders approached them for assistance. The approach of the Prime Minister of the MPR helps to construct a historical narrative that proves the United States did not compel Mongolia to embrace democracy and capitalism; instead, Mongolia asked for it. The Mongolian Deputy Prime Minister's visit to Washington unreservedly opened the gate for high-profile visitors from the United States; it is worth mentioning just a few from the 1990s: US Secretary of State James Baker in August 1990;

Secretary of State Madeleine Albright in May 1998; First Lady Hillary Clinton in September 1995, and again in 2012 as Secretary of State (Addleton 2013, 44, 46, 47), and many more. Mongolia has made efforts to seek assistance from third neighbours since the 1980s; consequently, Mongolia was not just an 'oasis of democracy', as US Secretary of State John Kerry claimed during his visit in June 2016 (Torbati 2016): it was also an oasis of global capitalism.

The above narrative of Byambasuren about the establishment of capitalism in Mongolia is unconventional when considered alongside some literature of capitalism (Rossabi 2005), which delineates how the United States and the West hook and entrap Third World countries to expand global capitalism to increase their surplus. To be sure, some Mongolian rulers searched for an alternative social, economic and political structure. Yet it is important to also recognise that the third neighbour policy met the American interest in expanding capitalism. In what follows, I intend to demonstrate that capitalist markets have been present in Mongolia since the 1990s, not only due to the influence of the United States but also US-based international organisations. For example, David Graeber writes about how the IMF and Western banks started the Third World debt crisis:

> During the '70s oil crisis, OPEC countries ended up pouring so much of their newfound riches into Western banks that the banks couldn't figure out where to invest the money; how Citibank and Chase, therefore, began sending agents around the world trying to convince Third World dictators and politicians to take out loans (at the time, this was called "go-go banking"); how they started out at extremely low rates of interest that almost immediately skyrocketed to 20 percent or so due to tight US money policies in the early '80s; how during the '80s and '90s, this led to the Third World debt crisis; how the IMF then stepped in to insist that, in order to obtain refinancing, poor countries would be obliged to abandon price supports on basic foodstuffs, or even policies of keeping strategic food reserves, and abandon free health care and free education; how all of this had led to the collapse of all the most basic supports for some of the poorest and most vulnerable people on earth. (Graeber 2011, 2)

Similarly, David Harvey (2010, 26–9) narrates the history of capital accumulation from the 1750s and points out that capitalist accumulation constantly requires that new profitable outlets be found. Global profit margins began to fall after a brief revival in the 1980s. Therefore, 'in a desperate attempt to find more places to put the surplus capital, a vast wave of privatisation swept around the world carried on the backs of the dogma that state-run enterprises are by definition inefficient and lax and that the only way to improve their performance is to pass them over to the private sector' (Harvey 2010, 28). This process can be seen in the targeting of the dismantling Soviet Bloc, including Mongolia. Mongolia did not single-handedly make the form of capitalism that exists within its borders; instead, it has been a by-product of local and global processes (Tsing 2005). In the words of Timothy Mitchell (2002), it is the 'rule of experts', or more precisely the rule of Mongolian and foreign experts that play an important role in the making of capitalism in Mongolia. Morris Rossabi (2005, 38) shares this approach and notes that The Asia Foundation (TAF) was the first US private institution to respond to Mongolia, launching its first programmes in 1990 and establishing its office 1991. The IMF and the Asian Development Bank (ADB) sent groups to study the Mongolian economy and to interview Mongolian pioneers of capitalism – namely, D. Ganbold, economist and first deputy minister – and other like-minded economists. An IMF research team conducted an official visit in August 1990, and ADB staff arrived in Mongolia in late May 1991 (Rossabi 2005, 43).[6] Consequently, as Lhamsuren Munkh-Erdene writes, 'Under the supervision of the International Monetary Fund, the World Bank and the Asian Development Bank, Mongolia's neophyte free marketers zealously launched a shock therapy (or structural adjustment) program in 1991 to establish a free market economy' (Munkh-Erdene 2012, 63). As such, in the next two decades, Mongolia's opening up to the West immediately flooded the country with a wide range of political, financial, trade, development, environment, human rights, religious, and philanthropic international donor and aid organisations, and trans-national corporations and investors from all around the world, all of which contributed to the making of capitalism in Mongolia.

Reflecting on the situations discussed above, Byambasuren notes that the Mongolian government had three critical tasks to urgently complete. First, Mongolia had to consolidate its independence, because many important decisions about Mongolia were previously made in Moscow. Second, Mongolia had to properly eliminate socialism – that is, it could not remain a relic of a socialist society. This was also important in order to demonstrate to the United States that Mongolia would do more than merely change the colour of socialism while continuing to follow Russia or China. Third, it was essential to expand Mongolia's economic connections to the world beyond the Soviet Bloc and China. In brief, for

Byambasuren, ending the authority of the Soviet Union in Mongolia and escaping possible Chinese integration was a top priority for democratic reform. Overall, Byambasuren argues that establishing and intensifying the relationship between the MPR and the United States was the best political move for the MPR to ensure the end of socialism and to make space in the region for the influence of third neighbours as powers to balance Russian and Chinese domination. Here, then, anxieties over the loss of independence were central to the direction in which the Mongolian economy developed following the end of socialism.

The National Economy

Similar to its independence, Mongolia's national economy was precarious, enabling political leaders to reify the national economy and justify their decisions (see also Munkherdene 2018, 375) to develop the mining industry in order to fund the emerging sovereign nation-state of Mongolia. For instance, for Byambasuren, embracing the United States was the best political move to properly end socialism and to terminate the domination of the Soviet Union and halt the PRC's attempt to influence the MPR. However, for Ochirbat Punsalmaa, a former mining engineer, who served as deputy Minister of Mining and Geology (1972), deputy of the People's Great Khural (1976), a member of the MPRP Central Committee, Minister of Foreign Economic Relations and Supplies (1987), and the first President of Mongolia (1990–97), embracing the liberal economy and free market principles of capitalism were political tactics to entice third neighbours and international donor organisations to fund the nation-state and to consolidate political independence. As such, the process of market liberalisation, privatisation and resource extraction contributed to anxieties over the loss of independence. It is to Ochirbat's account that I now turn to further illustrate.

When I called Ochirbat to set up an appointment, I explained the purpose and importance of my work. Towards the end of our telephone conversation, Ochirbat spoke about his meeting with Queen Elizabeth II during his presidency. He revealed that the Queen had asked him how Mongolia had managed to survive as an independent state in the face of the dominance of its two gigantic neighbours. Ochirbat, rather than revealing his reply, said he wanted me to seek an answer to this question through my research. During this telephone conversation, Ochirbat expressed a similar fear of the loss of independence that Byambasuren expressed, which I term the anxiety of independence. When I met him in his office at the Mongolian University of Science and Technology the following day, Ochirbat told me that in the aftermath of the fall of socialism, Mongolia had to embrace democracy, the free market, and a liberal economy not just to fund the emerging nation-state but also to secure the independence of the country by attracting third neighbours.

To assist the collapsing economy, Ochirbat said that he had initiated the Gold Programme (*Alt khötölbör*) in 1991. The first Gold Programme started in 1992 and ended in 1999 with success. The programme started its continuation in 2000 but encountered heavy resistance and stopped. The second Gold Programme started in 2017 and is planned to continue till 2020. Ochirbat explained that in the early 1990s in a situation with no currency reserve, export, investment or capacity to repay loans (see also Ichinkhorloo 2018, 391), and with inflation at 325 per cent, it had been impossible for the country to secure loans and attract investment (see also Byambajav 2015, 93). As such, he justified his actions by arguing that the quickest and easiest way to assist the economy was to exploit gold deposits (see also Bold 2013). Later in 2014, writing about such prgrammes to assist the national economy, Rebecca Empson and Tristan Webb explained that 'the economic plan of Mongolia's government requires foreign investment from private international investors for three main reasons: first, to develop the country's export-earning potential; second, to bring in world-class technology and know-how that will allow development of an indigenous capability in this field; and finally, to advance Mongolia's economic and political independence from Russia and China. The country has defined all of these areas as important for the country's national security' (Empson and Webb 2014, 241).

Before meeting Ochirbat, I spent time with his colleague Algaa Namgar, a metallurgical engineer and the director of the Mongolian National Mining Association (MNMA) founded by Ochirbat and his colleagues in 1994. In the early 1990s, Algaa worked for the Government Agency for Mining and his duty was to implement the Gold Programme. For Algaa, at the start, the Gold Programme was a dream (*möröödöl*). They had to create an attractive political, legal and socio-economic environment to appeal to foreign direct investment. Algaa explains that the World Bank supported Mongolia in trying to achieve the aims of the Gold Programme. The country also carefully followed the guidelines of the World Bank as it took steps to develop its mining economy. In 1991, the World Bank researched Mongolia in order to provide technical assistance for the development of mining. The project report produced recommendations on the geological potential of mineral resources, the capacities of the existing mines, the legal environment, and the government agencies and institutions needed to manage and assist the mining sector. Algaa and his colleagues' main project at the Government Agency for Mining was to draft a new Minerals Law following the recommendation of the World Bank. In 1994, the communist party won the first democratic election and complained that the bill was too liberal and rejected it (see also Chapter 2). As a result, the parliament supported a different version of the Minerals Law and passed it in 1994. Ochirbat said that many of the potential investors he met during his presidency complained that the 1994 Mongolian Minerals Law did not give enough clarity to the country's intention to collaborate with foreign companies and could even be read as suggesting that Mongolia would not cooperate. According to Ochirbat, the failure of the 1994 Minerals Law to appeal to investors forced the government to attempt to draft a different bill, again with the help of the World Bank. This time, it was not Algaa but another member of the MNMA, Jargalsaikhan Dugar,[7] who held a key position in the process to draft the new bill and to collaborate with the World Bank. I first met Jargalsaikhan in September 2015, at the 13th MNMA annual investment forum in Ulaanbaatar, titled 'For mining without populism', where he was the main organiser of the event. When I met him again in April 2018, he explained to me that the 1994 Minerals Law was an experimental process that revealed the importance of improving the law to make it more appealing to investors (which is the same point that Ochirbat had made).

The failure of the 1994 Minerals Law to appeal to the investors became the justification for Ochirbat, Algaa and Jargalsaikhan to further promote the liberal version of the Minerals Law. Jargalsaikhan explained that behind the liberalisation of the resource economy was not simply the will of the World Bank, as many Mongolians often complain (Sanchir 2016). It was, instead, a by-product of what Mongolian policymakers learned from their experiences. It was after the Mongolian Democratic Union coalition won the election in 1996 that the goal of fully liberalising the resource economy was accomplished. This time it was not only the president but also Prime Minister Enkhsaikhan Mendsaikhan who supported the law and pioneered the new mining economy. By winning the election, the democratic and market reformers therefore became another major state-driven force behind the passing of the 1997 Minerals Law, known as the 'liberal law'. Dalaibuyan Byambajav (2015, 93) writes that 'the new Mineral Law of 1997 was praised by the international mining community as one of the most liberal mining laws in Asia: it significantly relaxed rules for obtaining a licence and permitted full foreign ownership of mining ventures'. According to Algaa, one of the crucial changes in the law was the liberalisation of mining and natural resources and the stability agreement with mining companies.

Consequently, starting from 1997, the law permitted the government to issue thousands of exploration licences for almost no cost (Bulag 2017, 132), except for a small amount of administrative and registration fees and taxes. According to the Mineral Resources and Petroleum Authority of Mongolia (MRPAM), as of 2015, approximately 3,329 mineral licences have been granted to both foreign and domestic companies covering 13.9 million hectares (or 8.9 per cent of the entire territory of the country), of which 1,494 are operational and 1,835 are exploration licences (Ganbold and Ali 2017, 4).

The 1997 Minerals Law successfully transformed Mongolia into an attractive country in which to engage in mining operations, drawing in mining companies from North America, Europe and Australia. Boroo Gold, an Australian mining company, was one of the first and most famous examples of the foreign mining companies working in Mongolia. Under the Stability Agreement signed with the government of Mongolia in 1998, Boroo Gold was exempted from corporate income tax: a 100 per cent exemption for three years from the start of commercial production in 2004 and 50 per cent exemption for the next three years after that. The World Bank, the supporter of this law, praised this move as 'one of the strongest legal presentations of mineral licensee rights and obligations in the world, and the most investor-friendly and enabling law in Asia' (World Bank 2004, 52). Some Mongolian economists, namely Khashchuluun Chuluundorj and Enkhjargal Dandinbazar, note that the liberal economic regime for investments in the mining sector resulted in a rapid increase in the inflow of foreign investment in the natural resource sectors; Mongolia became one of the top 10 destinations in the world in regard to resource exploration investment (Chuluundorj and Dandinbazar 2014, 293). In a televised interview with Khashchuluun during which he discussed the case of Boroo Gold, he argued that this was an important step taken by the government of Mongolia to entice foreign investors, enlarge mining and support the national economy. In fact, there was a dramatic increase in the annual production of gold from 4.5 to 10.2 tonnes from 1992 to 2000, and from 11.8 to 24.1 tonnes from 2000 to 2005, which created about 10,000 jobs, and made significant contribution to the GDP, up to 20 per cent (Bold 2013; Mineral Resources and Petroleum Authority of Mongolia 2015, 7–8).

In brief, the above narrations of people who had important positions in the building of the emerging nation-state show how different discourses – about the critical situation of the economy; regional politics of Russia, China and Mongolia; prospective influence of the so-called third neighbours; opportunities of the gold mining – contribute to the reification of an imagined entity that can be called the national economy. The first President Ochirbat and his colleagues present these narratives as a grand story of the national economy to explain and justify why neoliberal policies that established capitalism in Mongolia were inevitable. The manner in which the national economy was central to anxieties surrounding independence can be seen via examining the responses of the above figures to criticisms and general discontent regarding the direction of economic and mining liberalisation found in the population at large.

Later, many Mongolians, including politicians, technocrats and activists, heavily criticised the case of the Boroo Gold and condemned those who established the liberal mining economy and invited foreign investors. For instance, they complained that land in the form of mining licences was distributed to private mining companies and caused significant damage to the environment (see Chapter 2). When I brought such criticisms up, Algaa responded that almost all of the licences were for mining exploration, not for extraction. He claims that mining exploration does not impact upon the land and the environment nearly as much as process of extraction. One of the final points that Algaa made was about the national economy. Those who possessed exploration licences had to pay an annual tax per hectare, which was an excellent contribution to the national economy. For these reasons, during the years when the economy of Mongolia was only just emerging, it was the best and most efficient way to use its vast territory for economic profit. In other words, soil and subsoil of Mongolia, with or without mineral resources, was entirely converted into a zone of economically efficient capitalist production in order to support the national economy. Similarly, Jargalsaikhan says that with the help of the 1997 Minerals Law, he initially wished to bring at least 1 per cent of the total investment in global mining exploration to Mongolia. But according to him, the law actually managed to bring in 5 per cent of the total global investment, which provided an enormous contribution to the Mongolian economy.

There is another common complaint by many Mongolians regarding the Boroo Gold mining company. The company stands as an ultimate example of how foreign companies exploit Mongolia and how mining can be *not* profitable to Mongolia. Many people also complain that the company was not taxed enough (see Chapter 2). Ochirbat responds to such accusations by stating that the Boroo Gold project was the first significant Western foreign direct investment that helped the

Mongolian government make its resource sector appealing to foreign investors and to establish trust needed for mining companies to operate in Mongolia. For Ochirbat, this was the success of the liberal Minerals Law from 1997. Since there were no previous instances of Western mining companies operating in Mongolia, and the 1994 Minerals Law failed to appeal to investors, Mongolian policymakers in the mid 1990s had to create the best possible social, economic, legal and political environments for foreign investors to operate mines in Mongolia in order to attract investors and to urgently bring growth to the national economy. He used a Mongolian proverb to describe the situation: 'dogs would not sniff when covered with fat; cows would not eat when covered in grass' (*öökhönd booson ch nokhoi tooj shinshlekhgüi, övsönd booson ch ükher tooj shinshlekhgüi*). In the case of Mongolia's mining sector's political and economic environment, the liberal Minerals Law was the 'grass' or 'fat' to entice investors. Therefore, he considers the Boroo Gold case as a project to entice foreign direct investment (FDI) and build trust, which appears to have been broadly successful.

The rule of President Ochirbat (1990–97) and his collaboration with the Mongolian Democratic Union coalition government (1996–2000) successfully established what Stuart Kirsch (2014) calls 'mining capitalism' in Mongolia. But unlike Kirsch, my focus in this chapter is not the conflict between transnational mining corporations and local communities conflict (Kirsch 2014, 9–14), but on how mining capitalism was embraced by Mongolian political leaders to fund the nation-state and to manage its independence.[8] The initial purpose, as Byambasuren and Ochirbat explained, was to urgently find a solution in the precarious situation to support the politically and economically devastated post-Soviet nation-state of Mongolia. For them, the available solution was to be found in the free market, liberal economy and global capitalism. However, both Byambasuren and Ochirbat claim that they had intentions to build a *khümüünleg, ardchilsan, irgenii niigem* (humane, democratic and civil society) and declared it in the 1992 constitution. An older version of this imagined society first appeared in the 1960 constitution *khümüünleg, ardchilsan, sotsialist niigem* (humane, democratic and socialist society). Finally, in the 1990s, the Mongolian state had a chance to make the imagined society real. This is also what many other influential democratic politicians claim – namely Zorig Sanjaasuren and Oyun Sanjaasuren, as documented by Morris Rossabi (2005, 34–5) – that political leaders of Mongolia had no particular intention to shift to capitalist markets. Instead, they intended a different society that is definable neither as socialism nor capitalism, but rather as a humane, democratic and civil society with the best possible advantages of both socialism and capitalism. In this way, the narrations of the above individuals demonstrate the agency and deliberate intentionality on the part of politicians and technocrats to build an economically and politically secure independent nation-state. In other words, for those political leaders, all of the above was done with nationalistic purpose to save the country. When I talked to Algaa, he said that although few people call what he and his colleagues did to preserve independence and stabilise the national economy 'nationalism', he argues that it should be considered as nationalist in the sense of *ündesnii üzel*.

In conclusion, the narrations by political figures I have provided show a number of alternative ways to think about the development of the Mongolian economy following the end of socialism. First, anxieties surrounding independence, the reification of the national economy, and attempts to secure political independence and bolster national economy in Mongolia led political leaders to influence the emerging nation-state in particular ways. In this process, the matter of sovereignty and the national economy became the priorities of the nation-state, or in the words of Appel, 'a privileged object – perhaps the privileged object – in official discourse' (2017, 294).[9] Second, the history of the Mongolian People's Republic, heavily influenced by the Soviet Union, consists of an entirely different story of the nation-state compared to postcolonial states in South America, Africa, South Asia and the Pacific. In other words, Mongolia does not have the same relationship that postcolonial nation-states have with the United States and other Western states. The absence of a colonial relationship enables Mongolian political leaders to create alliances with the United States and other Western states and embrace capitalism with considerably less coercion (see also Bumochir 2018b). This means that Mongolia, as a nation-state, is in a position that has a unique and unconventional relationship with those states. Third, taking advantage of the above historical background and justifications, those political leaders managed to bring the third neighbours (namely, America), international donors (such as the World Bank), and mining companies into the process of shaping the nation-state and building the modern nation. Fourth, the nationalist tendencies of the above-mentioned political leaders help me to explain the differences and disparities of forms of nationalism, which is also the focus of the next chapter. In this chapter, the nationalist tendency identifies precarity in the independence of the emerging nation-state by pointing out the weakness in the national economy emerging from the ruins of socialism. In contrast, 'resource nationalism', which I will explore in chapter two, also addresses issues in the political independence and sovereignty, not by reifying the national economy but by critiquing the lack of state control on land and territory, and by demanding fair distribution of mineral wealth, ownership and shares. The emergence of resource nationalism depicts a different political discourse, which supports state control in the resource economy in parliament and government, which was a response to many concerns, conflicts and resistances, and appeared as a result of the rapid growth of the gold mining.

Notes

1 Nationalism based on the reification of the national economy was a sentiment that was established in the MPR. In the Soviet-influenced style of planned economies, the notion of a national economy was central, and thus was the beginning of the *reification* of the national economy in Mongolia that this chapter describes. Following the collapse of socialism, many Mongolian politicians continued this form of nationalism and infused it with the issue of independence. But this chapter does not intend to trace the origin of the reification and nationalism of the national economy and much work needs to be done to show it.

2 For instance, in 1915, as a result of the tripartite conference between Mongolia, China and Russia in Khyakhta on the north border, China and Russia imposed their decision to keep Mongolia as an autonomous region of China. Consequently, Chinese troops occupied Mongolia in 1919 (Bulag 1998, 12; Atwood 2004, 91; Batsaikhan 2007). Also, in February 1945, just after the Second World War, Joseph Stalin negotiated Mongolia's independence with the United Kingdom and the United States at the Yalta Conference. Here, it is essential to address that the United Kingdom and the United States persuaded China to recognise the freedom of Mongolia. Christopher Atwood (2004, 92) notes that America's influence on China helped the Nationalist Party to recognise a 'high-level autonomy' for Mongolia and Tibet.

3 Personal communication with Munkh-Erdene Lhamsuren, May 2018.

4 Many Mongolians, including historians and politicians, might respond to Munkh-Erdene's and Bolor's claims by arguing that there are many historical facts that demonstrate Mongolia's struggle to gain political independence; therefore, it is based in reality. For example, after the collapse of the Qing Dynasty, the government of Tsarist Russia did not recognise Mongolia's independence. It instead allied with the government of the Republic of China and accepted their renewed right to deny the recognition of Mongolia's independence as a legally reconstituted successor state to the Qing Dynasty (1636–1912) (Atwood 2004, 91; Bulag 2012, 1). As a result, this generated the impression that Mongolia was an autonomous state under Chinese suzerainty from around 1910 to 1921 – at least according to Tsarist Russia if not for Mongolia (see also Bulag 1998, 12; Atwood 2004, 91; Batsaikhan 2007).

5 The same thing also happened in the early 1920s. Alicia Campi and Ragchaa Baasan write that Prime Minister Bodoo Dogsom (1921–22) 'wanted the United States, in conjunction with Soviet Russia, to act as Mongolia's protector, especially in regards to negotiations with China' (2009, 105; see also Bulag 1998, 13).

6 Many scholars made similar commentaries, such as L. Munkh-Erdene: 'Under the supervision of the International Monetary Fund, the World Bank and the Asian Development Bank, Mongolia's neophyte free marketers zealously launched a shock therapy (or structural adjustment) program in 1991 to establish a free market economy' (2012, 63).

7 Jargalsaikhan is a mining economist who studied in the Soviet Union and trained in the United States. Since the 1980s, he has held various government positions, including Officer of the Ministry of Geology and Mining, Head of the Mining Department at the Ministry of Heavy Industry, Vice President of the state-owned extraction company Mongol Erdene Holding, Chairman of the Mineral Resources Authority of Mongolia, and Head of the Geology and Mines Department of the Ministry of Trade and Industry.

8 Not only in 1990 but also in 1911 – when the theocratic government of Mongolia proclaimed political independence – mining and particularly gold mining was immediately adopted to fund the emerging nation-state and to promote its national economy (Tuya and Battomor 2012; Bonilla 2016; Jackson and Dear 2016; Bumochir 2018b).

9 My next chapter shows how this object becomes less privileged in the contest against other objects in the process of building the nation.

2
Beyond 'Resource Nationalism'

> A nasty bout of resource nationalism in Central Asia is worrying investors brave enough to invest in frontier markets. Mongolia and Kyrgyzstan are at it, with the Kyrgyz government this week announcing it has revoked 46 gold mining licences in what it calls an attempt to clean up the mining industry. At least none of the Kyrgyz licences are for the country's major mining operations. The situation is different in Mongolia, which on April 16 suddenly suspended the mining licences of South Gobi Resources. (Watson 2012)

On 26 April 2012, the *Financial Times* published an article titled 'Mongolia: Mine ownership gets political', in which moves to restrict Mongolian resources to be solely available to national actors (that is, the suspension of international mining licences) were declared to be 'a nasty bout of resource nationalism'. In the following years, many similar accusations emerged that painted Mongolians and the Mongolian state as 'resource nationalist' (e.g., Stanway and Edwards 2012; Els 2012; Cashell 2015; Genota 2017; Venzon 2018). For example, Jean-Sebastien Jacques, CEO of Rio Tinto, one of the world's largest Anglo-Australian metals and mining corporation that owns 66 per cent of the Oyu Tolgoi (Turquoise Hill) gold and copper mine in Mongolia. In May 2018, after 'the government ordered Turquoise Hill Resources, a subsidiary of . . . Rio Tinto, to pay $155 million in back taxes', Jacques told investors at a conference: 'A significant industry issue is resource nationalism' (Venzon 2018). This rhetoric was amplified by the international press as well as some scholars. For example, following the classification of 'resource nationalism' introduced by Ian Bremmer and Robert Johnston (2009), Misheelt Ganbold and Saleem Ali argue that the 'resource nationalism' in Mongolia is 'revolutionary', which 'is often expressed as public unrest that demands the transfer for natural resources from private owners to public coffers' (Ganbold and Ali 2017, 1, 10). According to them, the impact of the type of 'resource nationalism' in Mongolia 'tends to hinder profitability of the mineral projects' and can create 'a serious lack of confidence among investors' and possibly a 'diminution in foreign direct investment' (10).

On the other side, many scholars reject the above use of the term 'resource nationalism'. Rebecca Empson and Tristan Webb (2014), Jargalsaikhan Sanchir (2016), and Julian Dierkes (2016) all use the term 'resource nationalism' in quotes. Empson and Webb write that 'resource nationalism' is an 'accusation' (2014, 241, 247), while Sanchir and Dierkes find that the use of the term is a 'designated' (*songomol*) (Sanchir 2016, 55) 'pressure' (Dierkes 2016) tactic of transnational corporations and investors designed to influence the nation-state. Similarly, Orhon Myadar and Sara Jackson point out that 'resource nationalism' is often 'used by those who promote neoliberalism and the open market as a pejorative label to silence public grievances' (2018, 1). Some of these authors argue that it is not a nationalism but attempts to 'balance competing responsibilities' Empson and Webb (2014, 247) or to 'control natural resources' (Sanchir 2016, 55). Empson and Webb (2014, 247) suggest that 'the charge of "resource nationalism" of Mongolia is perhaps an oversimplification of the drivers behind macro-economic and political decision-making in Mongolia. An alternative view of the "resource nationalism" accusation would be not, perhaps, a nationalist intent *per se*, but rather the outcome of the Government's having to balance competing responsibilities across a spectrum of partnerships.' To avoid such over-simplifications, Sanchir translates 'resource nationalism' to *bayalgiin ündesnii khyanalt* in Mongolian, which means 'national resource control'. In its avoidance of an under-considered conflation of 'resource control' and 'nationalism', Sanchir's translation and explanation of 'resource nationalism' is a sensible one. Therefore, I borrow his translation in this chapter to reveal the richness and complexity of matters that are often simplified by the use of the pejorative label 'resource nationalism'.

The problems found with the term 'resource nationalism' can also be found elsewhere. John Childs (2016) and Natalie Koch and Tom Perreault (2018) problematise the neoliberal bias in the use of the term 'resource nationalism' and critique the narrowly defined reductionist framework of the concept in the fields of international relations, political science, applied economics, and business as a problem created more for neoliberal ideology than reality. Childs (2016, 530) notes that 'the definitions of resource nationalism that follows is highly divergent, often contingent on political interests and conceptual biases' and 'it is also constructed in a number of different ways'. Then he warns that 'resource nationalism should be not read as the simple opposite of familiar neoliberal imperatives of resource governance but as something which is always hybrid and in flux' (2016, 530). Child's account of 'resource nationalism' as 'always hybrid and in flux' is a description of a concept, rather than an existing phenomenon. It is an accusation levelled at various parties at different times, so the exact content of the term is bound to change. For this reason, Koch and Perreault (2018, 15) suggest a 'critical' approach to 'push debates beyond essentialist market- and state-based analyses of resource nationalism and to provide a far more nuanced approach to its various manifestations'. Taking this call into consideration, this chapter attempts to provide a far more nuanced approach to the concept as it manifests in Mongolia.

Following the above arguments developed by Childs and others, here I will treat 'resource nationalism' as a label and rhetorical or discursive device that describes some political processes and tactics aiming to control natural resources and to ensure Mongolia and its citizens receive their material benefits. The research materials that follow – which concern the resource economy in Mongolia – reveal a diversity of issues that lurks underneath the 'resource nationalism' tag which cannot be detached from different cultural and historical aspects that are classically associated with nationalism.[1] Considering those historical constructions and culturally specific ideas, such positions are not a 'nasty bout of resource nationalism'; instead, they are predictable outcomes. My focus upon nationalist sentiments in phenomena described as 'resource nationalism' is different from those of Empson and Webb (2014), Sanchir (2016) and Dierkes (2016) who, as I mentioned before, do not explicitly examine nationalism in their discussion of 'resource nationalism'. Rather, following Childs (2016) as well as Koch and Perreault (2018), my rejection of the use of the term 'resource nationalism' does not exclude an account of nationalist sentiments in the discourse and movements serving to control and protect natural resources.

Cultural and historical influences are often ignored in the existing literature on 'resource nationalism'. Only few works mention the importance of such cultural factors. Koch and Perreault (2018, 2) note that 'resource nationalism takes both political economic and cultural symbolic form, often in ways that are interwoven and mutually reinforcing'. In the case of Mongolia, Empson and Webb (2014, 233–4, 239) and Ganbold and Ali (2016, 10–11) underline the importance of the cultural factors and provide some illustrations on how the 'protective culture' of mobile pastoralists in regards to their land and natural resources shapes this so-called 'resource nationalism'. This chapter advances such an understanding of the protective culture of natural resources by shedding light on the (1) alternative cultural economy, (2) historical construction of the state control and protection of natural resources, and (3) political acts to balance contesting responsibilities and concerns. Thus, the following sections will explain why and how the Mongolian state and people promote national resource control.

The Alternative Economy

Common Mongolian ways of thinking about the economy and its connection to politics might be seen as an alternative understanding of the economy. In this section, a school of thought of those described as 'resource nationalists' rethink 'the economy' and construct an alternative understanding of the economy using two concepts. The first one stems from strategies from the Soviet practice of planned economies. The second one arises from ideas about consumption in Buddhism.

The key person to have pioneered the development of a so-called 'resource nationalism' in Mongolia is Khurts Choijin, a doctor of philosophy in geology and the former minister (1976–80) and vice-minister (1964–76) of Geology and Mining Productions (*Geologi, uul uurkhain üildveriin yam*) in the 1960s and 1970s. In addition to his positions in the ministry, he was also a member of the Mongolian People's Revolutionary Party Central Committee and of the People's Great Khural. It was not a surprise that the current President Battulga Khaltmaa, who is known to have nationalist sympathies (Dierkes and Jargalsaikhan 2017), awarded him the title *Khödölmöriin Baatar* (Hero of Labour) in December 2017. The award was in recognition of Khurts's achievements in the mining sector and his nationalistic position on resource policy. Not long after, in July 2018, a popular television documentary *Mongol tulgatany zuun erkhem* (Hundred distinguished figures of the Mongol hearth) – which President Battulga often supports – featured Khurts and his nationalistic views.

Khurts shares ideas and approaches with some experts from the National University of Mongolia, namely S. Avirmed,[2] a mining engineer and economist, and J. Byambaa, a geologist who closely collaborated with them to protest against certain neoliberal policies. Algaa Namgar, director of the Mongolian National Mining Association that I referred to in the previous chapter, refers to them as technocrats (in the following I will refer to them as nationalistic technocrats), based on their backgrounds and affiliations, a term that I will also use to describe Khurts and his colleagues. In 1994, Khurts was the key person who resisted the first liberal version of the Minerals Law that Algaa and his colleagues drafted and promoted with the help of the World Bank experts (see Chapter 1). At that time, the liberal version was not the only version of the bill. Khurts was in the working group to draft a different version of the bill that was supported and approved in 1994 (see also Konagaya and Lkhagvasuren 2014, 173–4). In other words, Mongolia already had a strong force of nationalistic technocrats: politicians who resisted the free market and liberal mining economy and instead supported state control, which is effectively the counter to the narrative I presented in Chapter 1. Not long after in 1997, those who promoted the liberal version of the law were successful in securing its passage.

In April 2018, I discovered that Khurts has an office at the Ministry of Mining and Heavy Industry. Khurts is in his 80s, and is retired. Eager to find out the precise nature of his role at the ministry, I went to see him one afternoon and found his office on the third floor of the ministry. Written on the door was *Mongolyn erdes bayalgiin salbaryn akhmadyn negdsen kholboo* (United Committee of Elders of Mongolia's Mineral Wealth Sector). There were two names under the label: 'Ch. Khurts and Ts. Baljinnyam'. Unfortunately, the room was locked. In the corridor, I met a man who works in the ministry and asked him about this 'committee of elders'. He said that Khurts comes almost every day and stays until around three or four in the afternoon. The man also told me that they are *chölöötei khümüüs* (free people), probably in comparison to people like himself, who do not have the same freedom to leave the office during working hours or to freely critique the government as these elders do. Later, Khurts claimed that their committee office and the employment in the ministry is largely symbolic (*belgedeliin chanartai*),[3] and the younger generation of policymakers do not take their opinions seriously. The next morning, I met Khurts in his office and visited him again several times over the following days. Each day I met him, visitors came one after another to see him and to congratulate him for receiving the distinction from the president. Almost all of the visitors were from the mining and geology sectors. Some brought the ceremonial silk scarf *khadag*, a silver bowl with milk, or gifts such as expensive alcoholic drinks. This is a common way to show respect in Mongolia. Some of the visitors would sit and talk to him for hours, while I served those visitors tea, coffee and biscuits. Their conversations were about Mongolia's imperial history, nomadic culture, socialism, geology and mining.

From these conversations, and also from the interviews I conducted with him in the following days, I learned that Khurts deeply regrets that the so-called neoliberal transformation – with the help of the World Bank – demolished, in his eyes, Mongolia's economic, political and scientific mineral resources institutions, which were developed during socialism under his direct leadership. He talked about this intensively with the people who came to visit him in his office, including myself. He told me that the market and democratic transformation made his 4,500 engineers unemployed and turned them into *naimaachin* (suitcase traders)[4] and closed down his research institute with over 100 researchers, which was a significant loss for the country. He complained that all of the geology and mining institutions were dismantled and replaced with apparatuses of the free market economy and neoliberalism. Regarding this he said, '*chikagogiin malchikuud, Milton Friedman, dendüü muukhai, khudlaa, khooson surgaali gargaj irsen*' (Chicago boys such as Milton Friedman developed an awfully false and empty teaching). He claimed 'there is no economy without the state' (*törgüi ediin zasag gej baikhgüi*) (see also Callon 1998; Çaliskan and Callon 2009; Appel 2017, 301). I asked him why he thought neoliberal economic theory was false. In answer to this, he told me how he became a minister and a politician.

He was an outstanding student in Moscow during socialism, and one of the first to return to the Mongolian People's Republic (MPR) with expert knowledge on geology. Upon his return to Mongolia in 1963, the National University of Mongolia and the Government Agency for Geological Exploration and Research immediately employed him. Not long after, he was appointed as the director of the Sector of the Economy and Planning in the Agency for Exploration and Research, where he began studying the economy. To re-educate himself he had to read many Soviet-era Russian books about the economy, planning and strategy. These texts claimed that all countries (*uls oron*) must have a centralised strategy (*strategi*) and policy (*bodlogo*) to exist. Moreover, he explained that such strategies help the state (*tör*) to complete its ultimate duty to protect its people, territory, environment and culture. Most importantly, different political strategies do not let any sectors – for instance, the mineral sector or the economic sector – remain free from national politics.

For Khurts, this necessity creates a level of state participation and control in one way or another (see also Polanyi 2001), which means that the idea of the free market and liberal economy is a false dream for him.

He reached such conclusions from his experience working as the minister of Geology and Mining Productions. When he was a minister (1976–80), he had to make a strategic decision to not invest in the extraction of uranium and crystal used in optics. According to him, there were several problems regarding the extraction of these two minerals. The ministry was spending 28 per cent of its total budget for the exploration of crystal and 8 per cent for uranium mining, while they did not bring much profit to the GDP. The Soviet Union needed these minerals and Mongolia had to export them at almost no cost, in contrast to the international market price. He also complained that uranium mining was highly destructive to the local people and the environment, and it was one of the riskiest mining explorations. Concerning these two issues, Khurts had to make a strategic decision to stop investing in these two mineral explorations and to close down their operations. Consequently, 30 per cent of the Russian employees in the MPR's geology and mining industry had to return to the Soviet Union. Instead, Khurts recommended investing in building industries that process natural resources such as copper, in order to support the national economy. While clearly his policies were not resisting 'neoliberalism', the necessity for interventionist economic policy appears to have been solidified through this experience. Unfortunately, as a result of this decision and because he fought against the interests of the Soviet Union, Khurts eventually lost his position as minister.

According to the principles of the free market and liberal economy noted above, political strategies that assert control over the free market violate the rights of private companies. To respond to this, Khurts explains how the understanding of the economy in Mongolia is and should be principally different from the understanding of the economy in capitalism. First, he claims that the culture in Mongolia does not prioritise private property and individual rights but instead prioritises the common and commoners.[5] For him, the matter of the nation can never be considered as less important than the question of the individual or private companies. By prioritising the national economy and the issue of funding the emerging nation-state, he shares the same argument with Ochirbat Punsalmaa and others, which I presented in Chapter 1. The difference is Khurts had to close down mines, while Ochirbat had to open mines to support the national economy. For Khurts, the two are both strategic political decisions for the sake of the national economy. I must add that both are also nationalistic decisions, but the former (as seen in Chapter 1) is often considered in the terms of contemporary neoliberal ideology and Mongolian discourses as an example of *ündesnii üzel* (a good nationalism that supports prosperity and development), while the latter stands as an example of *ündserkheg üzel* (a bad nationalism that blocks development, prosperity and democracy) or 'resource nationalism'.

Moreover, Khurts elaborated on another principle that also makes some widespread Mongolian ideas of the economy different from those found in so-called liberal economies that justify the state strategic intervention in the economy in a different way. To explain this, he used teachings from Mongolian Buddhism. He learned about 'correct consumption, correct demand, and correct outcome' (*zöv khereglee, zöv kheregtsee, zöv ür dün*), which summarises a strand of Buddhist philosophy concerned with the economy (see also Brown 2015).[6] Khurts elaborated upon this in the following sense: 'Correct consumption' teaches us how to manage by efficiently using what we have. 'Correct demand', the opposite of unlimited demand, is about how to limit one's demand or greed (*shunal*) based on the efficiency of the 'correct consumption'. 'Correct outcome' is where the 'correct use' and the 'correct demand' meet. He believes that these sorts of ideas should be recognised in the way they have influenced conceptions of the economy in contemporary Mongolia. In principle, there is a significant difference between the certain understandings of the economy in Mongolia and capitalism that lies in the idea of greed (*shunal*): one constrains greed while the other, for Khurts, rewards greed. Validating greed in Buddhism is a sin that is discussed widely even in contemporary Mongolia. Just as the first principle, arising from the soviet planned economy, the second principle also justifies state planning, strategy and control of the economy. These two principles are important influences upon some of the ideas related to the economy that circulate in Mongolia to this day. Certainly, they are important in the manner in which they motivated Khurts to promote articles related to the state ownership of the minerals with strategic importance in the 1994 Minerals Law (Mongol Ulsyn Ikh Khural 1994, Article 4), and the state ownership of the deposits with similar importance in the 2006 Minerals Law (Mongol Ulsyn Ikh Khural 2006, Article 5). In this sense, the state control of natural resources can be rethought as a consequence of the specific ideological resources that circulate and are contested in contemporary Mongolia. The salience of these two principles is one of the reasons why many people such as Khurts support various forms of state control and protection.

Both of the principles mentioned above endorse forms of strategy, control and governance of the economy. Indeed, one might argue that this can be further seen in the very term *ediin zasag* which is used to translate 'the economy'. As I argue in a different paper (Plueckhahn and Bumochir 2018), the conceptualisations of economy in this term differ from understandings of 'economy' in English. It consists of two words. *Ed* means article, item, thing, property, possession, wealth and so on, mostly referring to something material. *Zasag* means governance, rule, or authority; its verb version, *zasakh*, means fix, do, make, manage, organise, control, rule, master or govern (Plueckhahn and Bumochir 2018, 347). As David Sneath (2002, 201) explains, *ediin zasag* literally means the 'governance of property' or 'possessions authority' and 'the very definition of the economic sphere depends upon the notion of political authority', which reveals a Mongolian linguistic interconnection between politics and the economy in the very act of speaking about it.

State Control and Protection

Varied historical narratives from different periods of the history of the nation of Mongolia make state control and protection of natural resources a normative discourse that is available for Khurts and other nationalist technocrats and politicians to deploy. This section introduces several historical narratives that depict a period of time starting from the time of Xiongnu Empire in the 3rd century BC–1st century AD to the Soviet regime in the twentieth century. As I find in this section, state control does not have a particular and stable meaning and a form. People give different meanings to state control and construe different forms of state control for certain purposes, usually to argue against neoliberal and free market ideologies, using historical narratives. The historically constructed validation and acceptance puts state control and protection of natural resources in a privileged position comparable to other privileged objects such as the national economy. Hanna Appel writes, 'In Equatorial Guinea, as elsewhere, the economy is a privileged object – perhaps the privileged object – in official discourse' (2017, 294). In contrast to Appel's argument, this section shows how historically constructed indigenous perspectives on state control and the protection of land and natural resources create alternative privileged objects besides the economy. As a result, there is a contest of privileged objects in Mongolia and, as I illustrated in the previous section, political rulers struggle to balance those different priorities. In my reading, the term 'resource nationalism' has come about without proper consideration of how nationalist sentiments labelled as such are related to historical constructions of state control over natural resources. For many Mongolians, all of the above-mentioned laws, regulations and decisions identified in the framework of 'resource nationalism' remain as diverse issues and problems that go far beyond the framework of the reductionist accusation this term conveys. In the discussion below, I will shed light on matters of 'tradition' and history that are unavoidable in Mongolia. As Ganbold and Ali note, 'the long-held nomadic belief and lifestyle of herders in Mongolia is an important factor for shaping such trends' (2017, 6–7). But unlike Ganbold and Ali, I argue that this should not apply to understanding nationalism in the reductionist framework of 'resource nationalism'. The effects of nomadism and the pastoral way of life upon the relationship of land and herders are not captured by the term 'resource nationalism', but are specific to nationalism in Mongolia. Here, I will not look at the nomadic and pastoralist culture *per se* as Ganbold and Ali mention, but I will focus on the historical construction of state control of natural resources which occurred *within* a pastoralist context.

Nationalistic technocrats, scholars, politicians and environmental movement leaders often deploy historical narratives to promote and justify state control of natural resources. According to many of these agents, and the existing literature on the history of mining in Mongolia, the tendency of the state to assert authority over resources existed in presocialist times. In Khurts's office, I found a thick book titled *Mining of Mongolia 95: Routes of the historical development of mining in Mongolia* (Magvanjav and Tsogtbaatar 2017) dedicated to the '95th anniversary of mining in Mongolia'. The first chapter of the book reminds readers of the oath of the Xiongnu emperor Modu Chanyu (234 BC–174 BC) that circulated during the domination of the Qing Empire (1644–1912) – *Burkhan guisan ch sööm gazar büü ög* (Do not give away an inch of land if even God asks for it) – which was often considered as an important message of resistance from the past (Magvanjav and Tsogtbaatar 2017, 24). The authors took this phrase for granted and provided no official references. According to some historical sources, this is not what Modu Chanyu actually said. In a myth about Modu, depicted by the ancient Chinese historian Sima Qian (135 or 145 BC–86 BC), his enemy first asked for his famous fast horse and then his favourite consort, which he agreed to give. Finally, the enemy asked for his land, and Modu flew into rage and said, 'Land is the basis of the nation' and 'he executed all the ministers who had advised him to do so' (Qian 1993, 135). In Mongolian, this is often translated as *gazar bol ulsyn ündes* (land is the basis of the people, nation or country; see also Kradin 2012, 54). Further, Mongolian historians claim that the Oirat ruler Galdan Khan (1644–97), who fought against the rule of the Qing Empire, is the one who said *Minii nutgiin gazar shoroonoos burkhan guisan ch bitgii ög* (do not give away my land even if God asks for it) (Dashnyam 2014, 228). It is common in Mongolia to mix those two phrases and quote them with no proper references.[7] In my interview with Khurts, in addition to mentioning these two phrases, he further suggested that the state control, regulation and protection of natural resources and the environment in Mongolia started in the times of the Mongol Empire (1206–1368), and it was explicitly declared in the Great Code of the Mongol Empire called *Ikh Zasag* (or *Yasa*, *Yasag* and *Jasag*).[8] Moreover, he went on to claim the advancement in production that stemmed from the successful regulation of natural resources contributed to the empire's conquests (see also Konagaya and Lkhagvasuren 2014, 158–9). Here, with the state control Khurts indicates state's or emperors' sovereign authority to use or to let someone use natural resources for certain purposes to help the state and people.

Another famous historical record that reveals state sway over natural resources comes from the mid-eighteenth century. Many of the books on the history of mining in Mongolia (see also Tuya and Battomor 2012; Sodbaatar 2013) proudly narrate this event. Under the rule of the Qing Empire, aristocratic nobles in Zasagt Khan and Sain Noyon Khan banners established and organised *altan-u qarayul* (a gold patrol) to protect wild animals, herbs, gold and all other natural resources from illegal exploitations and smuggling by migrant Chinese and Russians (Nasanbaljir 1964; Banzragch 2004; Tuya and Battomor 2012, 43–47; Jigmeddorj 2015; Magvanjav and Tsogtbaatar 2017, 24). I also had a chance to talk to Khurts about this history. He pointed out that gold patrol was one the most efficient ways to successfully protect natural resources from foreign threats in the history of Mongolia (see also Konagaya and Lkhagvasuren 2014, 161).

The first large-scale multinational mine in Mongolia under the Qing Empire was known as Mongolor. The company was financed by Russian, Belgian and Qing capital; it was also staffed with French engineers, American hydrologists, and Russian, Chinese and Mongolian miners. It became operational in March 1900 (Tuya and Battomor 2012, 68–70; Jackson and Dear 2016, 350; see also Dear 2014; Bonilla 2016). Many commentators have drawn an analogy between early resistance by Mongolian people against the Mongolor project and what is happening in contemporary Mongolia. Qing officials decided to transform Mongolia into a strategic buffer zone against the encroachment of the Russian Empire and into a profitable region through a combination of agricultural land reclamation and mining. The term *li yuan* (source of profit) was adopted to describe Mongolian soil (Sneath 2001; Jackson and Dear 2016, 349; Dear 2014, 245–7). Mongol aristocrats and civilians enacted strong resistance against mining operations (not only this company but mining operations in general) and the policy of the Qing authorities to exploit natural resources in Mongolia. Several important aspects may be deduced from the complaints and resistance against mining. Ya. Sodbaatar (2013, 31–3, 48, 53, 54, 55) has published extensive archival documents that decry Russian and Chinese gold and coal mining operations and the operation of Mongolor. According to these materials, many people complained about 'the incompatibility of mining to the way of living and the environment' (*aj törökh arga, oron nutagt kharshtai*). Most of those documents reveal complaints that suggest mining operations degrade land and pastures and violate local people's belief in land and water spirits. All of the charges described mining operations as foreign to Mongolian culture, not only as Russians, Chinese and other foreigners controlled these operations but also in the sense that they are destructive to the pastoral way of life and Buddhist and shamanic beliefs that teach not to destroy life of earth (Shimamura 2014, 395; see also Tanaka 2002).

Although mining turned out to be destructive and incompatible to this view of Mongolian culture, in 1911, when the theocratic state of Bogd Khan (1869–1924) established its government and proclaimed independence, the government immediately decided to nationalise mining for the sake of the national economy and its development. In other words, because mining turned out to be an opportunity to promote the national economy and sovereignty (Tuya and Battomor 2012; Bonilla 2016), the reification of the national economy became the justification to allow mining activities (Appel 2017). In 1913, the new theocratic government of Mongolia introduced its Mining Regulations, which revealed a desire to expand the mining industry by welcoming foreign companies (Batsaikhan 2009, 72; Tuya and Battomor 2012, 75–78; Sodbaatar 2013, 92; Bonilla 2016). Significantly, the new government continued with contracts with the existing mining companies and put efforts into opening further mining explorations (Tuya and Battomor 2012, 79; Sodbaatar 2013, 93). In the absence of the Qing Dynasty, Mongolor no longer had a responsibility to make payments to Beijing. Mongolia became the outright owner, thereby resulting in the state's access to revenues from gold mining (Bonilla 2016). As such, the rulers of Mongolia started to nationalise and commodify natural resources as valuable assets in the economy of the emerging state. Khurts pointed out that according to the contract made between Mongolor and the Mongolian government in 1908, Mongolia received 16.5 per cent of the total amount of gold extracted; this percentage later increased to 20 per cent (see also Magvanjav and Tsogtbaatar 2017, 7–8). He also claimed that the theocratic government of Mongolia used to own the deposits, and only issued permission to lease subsoil (see also Konagaya and Lkhagvasuren 2014, 163). For Khurts, this was again an example of how rulers of Mongolia promoted state control of natural resources.

The historical policies of state ownership, protection and control were re-enforced after the revolution in 1921. In the book I found in Khurts's office (*Mining in Mongolia*), I was eager to read the passages referring to 1922; that is, why that year was officially recognised as the year mining started in Mongolia. Khurts told me that in 1957, *Arad-un yeke qural* (People's Great Khural) issued a decision to celebrate 1922 as the start of the mining sector in Mongolia, after the revolution supported by the Russian Red Army in 1921. In December 1922, the first mine to be announced as a state-owned mine was a coal mine in Nalaikh. After two years, the first constitution of the MPR, passed in 1924, declared that natural resources are the property of people (and therefore the state) (*arad neyite-yin körüngge*). It should be noted that the constitution explains that 'from the past natural resources had been the wealth of the people and public' (*erte čag-ača inaγsi arad neyite-yin körüngge bayiγsaγar iregsen zang surγal*) (Mongγol ulus-un yeke qural 1924, Chapter 1, Article 1). This article of the constitution not only declares the ownership and authority of *ard niit* (people and public) but also pronounces that this is the moral custom and teaching (*zan surgaali*) that must be adopted and privileged by the next generation of Mongols. In other words, the constitution explains that it is essential to acknowledge, privilege and preserve Mongolian traditional customs and teachings (*zang surγal*) in regard to natural resources. As such, according to David Sneath (2010, 251), notions of tradition (*ulamjlal*) and historical narratives become resources for politicians to mobilise politically and to historically ensure state control, as noted in this chapter. Throughout the development of the socialist system, the Mongolian government further reinforced and validated the state's and people's ownership of natural resources. The 1940 (Bügüde nayiramdaqu mongγol arad ulus-un yeke qural 1940, Chapter 1, Article 5) and 1960 (Bügd Nairamdakh Mongol Ard Ulsyn Ardyn Ikh Khural 1960, Chapter 2, Article 10) constitutions of the MPR declare that natural resources are *ulsyn ömch ard tümnii khöröngö* ('wealth of the nation' or 'country' and 'property of the people'). Mette High (2012, 254) asserts that later laws, namely the Law on Subsoil (1988), the Constitution of Mongolia (1992), the Environmental Protection Law (1995) and the Law on Water (1995) emphasised state interests above those of individual mining investors even after the fall of socialism. For example, the 1992 constitution of Mongolia declares, 'except that given to the citizens of Mongolia for private ownership, the land, as well as the subsoil with its mineral resources, forests, water resources and wildfowl shall be the property of the State'. Moreover, the constitution also states that 'natural resources should be under the authority of people' (*ard tümnii medel*) and the 'protection of the state' (*töriin khamgaalalt*) (Bügd Nairamdakh Mongol Ard Ulsyn Ardyn Ikh Khural 1992, Article 6). In this sense, the anniversary of mining celebrates state ownership and the people's ownership of the territory and natural resources, as reinforced through reference to traditional customs and teachings (*zang suryal*).

As we have seen, the state control and protection of natural resources is a salient discourse that is available to deploy. Deployment of such constructions of the indigenous experience of history – and its political motivations – actively shapes the contemporary political economy of natural resources. Therefore, an account of mining in Mongolia, past and present, even when involving largely international companies, *cannot* be complete without recognising the availability, circulation and import that such *Mongolian* discourse has on the political economic context. By taking historical constructions into consideration, the state and the people of Mongolia do not simply control natural resources. They also attempt to protect the environment, locality and territory as well as the knowledge, beliefs, feelings and customs regarding land and localities they inhabit, which is a classic in romantic nationalist thought. These constructions also prompt 'people' (*ard*) to acknowledge state control instead of undermining it. Therefore, for many Mongolian policymakers and people, it is impossible to straightforwardly reject or even undermine the historical construction of state control and immediately establish an ultimately capitalist market-oriented economy.

State's Action to Balance

In the impasse in Mongolia's resource economy (see also Preface), the state attempts to reconcile competing concerns promoted in the schools of liberal and nationalist thought. To be more precise, political attempts to balance these schools of thought led to situations that can be called 'resource nationalist' by some people. In the words of Empson and Webb, the Mongolian rulers and politicians in parliament and government attempt to establish 'trusting partnerships' and 'to balance competing responsibilities' (Empson and Webb 2014, 247). 'The idea of "trusting relationship" can refer to that relationship between the Mongolian State and foreign investors, as well as in certain ways to that relationship between the State and the Mongolian people' (2014, 232). They go on to note that the Mongolian government 'cannot serve all of its relations', and this forces the government to put 'some "on hold" while attending to some others' (2014, 247). In this way, the government's (and parliament's) delicate attempt to balance contesting partnerships and their privileges sometimes puts the economy on hold and promotes state control of the resource economy. Such forms of political control of natural resources to maximise and distribute its benefits are often defined as 'resource nationalist' (Bremmer and Johnston 2009, 149; Wilson 2015, 399; Childs 2016, 539; Koch and Perreault 2018, 1), and mask the political struggles to balance competing interests that lay behind events.

For example, Algaa calls the 1997 Minerals Law a liberal law that successfully appealed to FDI;[9] Jargalsaikhan Dugar says that he was proud to promote the national economy by creating this law; and Ochirbat claims it helped to fund the emerging sovereign state. Yet many nationalistic technocrats, scholars and politicians such as Khurts rejected it and called it the 'black law' (*khar khuuli*). In my interview, Khurts gave a number of justifications for this. First, he claims that the law was written by a foreigner who was a World Bank consultant and who copied it from laws in English (Tsogzolmaa 2010) and that the foreigner did not consider the uniqueness of the Mongolian context. Second, as a result of the law, the territory of Mongolia was distributed in the form of mining licences to Mongolian and foreign private mining companies. Khurts claims that this violates an article in the constitution of Mongolia, which states: 'In Mongolia, the land, its subsoil, forests, water, fauna and flora and other natural resources shall be subject to people's power and State protection' (Bügd Nairamdakh Mongol Ard Ulsyn Ardyn Ikh Khural 1992, Article 6). Third, he complained about the *litsenziin naimaa* (selling of licences). As a result of the liberalisation in the economy, some people and companies obtained a large number of mining licences. He considers that this led to an unequal distribution of Mongolia's natural wealth. Fourth, he explained to me that the private ownership (*ömchlökh*) or the temporary possession (*ezemshikh*)[10] of land or natural resources (see also Sneath 2004; High 2012, 254; Empson and Webb 2014, 234 and 239) risks the above-mentioned duty of the state to protect and control. If the company owns or temporarily possesses the soil and the subsoil, then this opens a right for them to transfer and mortgage the soil and the subsoil they own or temporarily possess in the form of the mining licence, which was actually approved in the Minerals Laws in 1997 (Mongol Ulsyn Ikh Khural 1997, Article 40) and 2006 (Mongol Ulsyn Ikh

Khural 2006, Article 40). Khurts decries this and warns that if mining companies fail to repay their loans, the mortgaged soil and subsoil in the form of the mining licence will remain mortgaged to one of the foreign banks. He worries that in such a case, the Mongolian government will be obliged to pay off the debt of a private company in order not to lose its land. To prevent such risks, according to him, the soil and subsoil should be the property of the state and could only be leased to others (see also High 2012, 254). He successfully managed to implement this approach in the Law on Subsoil in 1988 (Bügd Nairamdakh Mongol Ard Ulsyn Ardyn Ikh Khural 1988, Article 1), and then in the Minerals Law in 1994 (Mongol Ulsyn Ikh Khural 1994, Article 33–7), but the Minerals Laws of 1997 and 2006 both eliminated his protection.

For the above reasons, Khurts and his colleagues fought against the 'black law'. This was true not only for Khurts and his colleagues; some other politicians and the public also expressed similar concerns. One of them was Enkhsaikhan Onomoo,[11] who became a parliament member in 2004. To support the nationalistic technocrats' group and to lobby the amendment, he founded a movement called *Minii mongolyn gazar shoroo* (My Mongolian Land and Earth). With the effort of Enkhsaikhan and a few other parliament members, the new Minerals Law was success-fully passed by parliament in 2006. The new law has some significant differences from the previous liberal version. For example, in the new version, Khurts and his colleagues managed to introduce articles concerning mineral deposits with strategic importance, and 34 per cent to 50 per cent state ownership (Mongol Ulsyn Ikh Khural 2006, Article 5). These articles also allocate a right and authority for the state to participate and control natural resources and the mining economy. There are also the following important changes: Chapter 2 of the new law (Mongol Ulsyn Ikh Khural 2006, Article 8–14) is entirely dedicated to the sovereign right (*büren erkh*) of parliament and the central and local governments to regulate (*zokhitsuulakh*) the resource economy; it eliminated the section to issue licences to those who first apply, and in its place introduced a stricter selection process and criteria (Mongol Ulsyn Ikh Khural 2006, Article 18–23, 24–6) and eliminated the section on the stability agreement in the new version. Here we see the influence of the discursive resources that encourage the state control of resources outlined throughout this chapter and the very real impact they have on the economy in Mongolia. Yet these amendments in the Minerals Law in 2006 marked only the start of the national resource control and the overall concern further echoed many more attempts of state control.

Another law that placed restraints upon the growth of the gold mining economy was the windfall profit tax law (*Genetiin ashigiin tatvaryn khuuli)*, which passed in 2006. In May 2006, parliament approved a windfall profit tax on copper and gold exports, which required companies to pay a fee at a rate of 68 per cent when the copper price exceeds US$2,600 per metric ton and the gold price reaches US$500 per troy ounce on the London Metal Exchange (Tse 2007, 1). As the government and parliament predicted from mining companies' tax, after the implementation of the law, the GDP increased by an average of 48 per cent from 2006 to 2008 (*Mongolian Mining Journal* 2008). For those who created and supported the bill, this success was the result of the windfall tax law. As such, the national economy appears to be the dominant justification for the legitimacy of the windfall tax law. For those who introduced the bill and those who supported it in parliament, this law was the best possible way for the government to claim its ownership of natural resources and to provide the most national benefit from the exploitation of natural resources. The high tax was also a response by some politicians to the complaints against those few Mongolian and foreign private mining companies who benefited the most from Mongolia's mineral wealth. As such, it is important to examine the story of how the government came to this idea in the first place.

In Mongolia, this law is also known as the law of Fortuna's daughter (*Fortunagiin okhiny khuuli*). Fortuna is the nickname of Batbayar Nyamjav, an economist and politician, who was a member of parliament and a minister of Construction and City Building. There are also rumours that he is a shaman. Batbayar started his private business in 1992 and established a company, which he called Fortuna. Later it became clear that the idea of the law came from his daughter, Jargalan Batbayar, who studied economics at Columbia University (2001–4). Many Mongolians make jokes by saying that she learned it on the internet or the idea came by way of Fortuna's dream. The origin of this law became an example for many people as to how politicians in Mongolia initiate and approve laws without serious research or knowledge. As the story goes, the law initially targeted copper, not gold, and particularly the Mongolian and Russian joint venture Erdenet Mining Corporation, with the intent to increase Mongolia's profits. According to Algaa's version of the story, it was unfair to target Erdenet and Russia only; therefore, seizing the opportunity of the gold price increase, the government decided to include both copper and gold and other gold mining companies in the law.

Unfortunately, in addition to the increase in the GDP, this law brought about a disastrous outcome in the gold mining sector, which I will also present in the next chapter on gold mining companies. Gold mining companies stopped selling their gold to the Mongol Bank, and the illegal trade of gold dramatically expanded in the years that followed. Official records indicate that the amount of gold production decreased from 24.1 tonnes in 2005 to 17.4 tonnes in 2007, and the export of gold dropped from 23.8 tonnes in 2005 to 11.5 in 2007. A *Mongolian Mining Journal* (2008) editorial calculated that during those two years, Mongolia lost about US$565 million from gold exports because of the windfall tax law and illegality it encouraged.

As a consequence of the above laws,[12] the production of gold further dropped to 5.7 tonnes in 2011 and gold export fell to 2.6 tonnes (Gold 2025 Programme baseline research report 2015, 7–8). The Bank of Mongolia purchased 15.23 tonnes of gold in 2005, which was an all-time high. This amount dramatically dropped to 2.12 tonnes in 2010, 3.31 tonnes in 2011, and 3.34 tonnes in 2012 (Bank of Mongolia 2017).

Since the 2000s, the boom of nationalistic sentiments and state control was not the only issue to generate the laws mentioned above. Bans constrained the growth of the small- and medium-scale gold mining operations in Mongolia. Further, the increase in the national economy, discovery of abundant minerals, and plans to open the world's largest mines to exploit deposits with strategic importance caused Mongolian rulers to undermine the operations of the small- and medium-scale gold mining companies. The new era of the vast mineral deposits extraction of transnational mining corporations such as Rio Tinto started to shape the gold mining sector in relation to small- and medium-scale companies. Rio Tinto's investment in the Oyu Tolgoi (OT) gold and copper mine made the Mongolian economy most invested in 2012 in the world by the IMF metric (Riley 2012). In 2011, Mongolian GDP growth rate reached an all-time high at 17.5 per cent. Compared to the 1990s, the national economy appeared to be in a much stronger state. In this sense, it was the start of the significant strategic mines that caused Mongolian rulers to cancel the windfall tax law, not the struggles and battles of the small and medium gold mining companies. To pave the way for the agreement of Rio Tinto and its partner Ivanhoe Mines to establish Oyu Tolgoi copper-gold mine, the government of Mongolia agreed to scrap the 68 per cent windfall profit tax on 25 August 2009. In preparation for the start of production, scheduled for 2013, the government rescinded the windfall law in early 2011 (Swire 2009).

The growth in the GDP rate and the presence of transnational corporations created a window of opportunity for some politicians to develop a different political campaign and a distinct nationalistic tendency, as we have seen in the above case of the windfall tax law. But this window of opportunity was temporary and brief, and the GDP growth rate collapsed again in 2012. From 2014, Mongolia's GDP growth rate dramatically dropped to -1.6 per cent in 2016 (Trading Economics 2018). At the end of 2014, Prime Minister Saikhanbileg Chimed of the Democratic Party finally admitted that Mongolia was faced with a severe economic crisis. The announcement of the crisis had a specific purpose: the government justified the reinforcement of the liberal economy against so-called 'resource nationalism'. As a result, in May 2015, Mongolia and Rio Tinto signed a US$5.4 billion deal to expand the underground development of Oyu Tolgoi. International media outlets quoted Saikhanbileg's statement: 'Mongolia is back to business' (*Guardian* 2015). In his speech, he also emphasised the importance of terminating all possible interventions of the Mongolian government and politicians in the business of the two companies. In many ways, this announcement was a redeclaration of Mongolia's free market economy. Saikhanbileg's appeal to terminate political intervention in mining was somewhat moot because the nation-state has a particular policy regarding strategic deposits and their wealth, which creates a situation where political interventions are unavoidable (Dulam 2015).

In 2016, the regime of the liberal resource economy was further enforced during the rule of the subsequent government of the People's Party. In 2017, the Mongolian government started 'Gold Programme Two', which is reminiscent of the first 'Gold Programme' in 1992. The purpose of Gold Programme Two was to liberate the resource economy, invite foreign investment, encourage activity in the gold mining industry, and to grow the GDP rate. Such political decisions made in 1992 and in 2017 show that the rise and decline of the two competing regimes were *both* tied to the reification of the national economy. The reification of the national economy has a dialogical relationship with the issue of the environment, social justice and distribution of natural resource wealth. When Mongolia achieves fast economic growth, then many rulers consider the environment, social justice and distribution of resource wealth; when the economy hits a crisis, many of them abandon the environmental problems and concerns about the resource economy. Yet in both cases, specifically *Mongolian* discursive resources related to the environment and economy shape the political-economic outcomes.

Notes

1 Jeffrey Wilson (2015) critiques the focus on the economy in the discussion of 'resource nationalism' and underlines the importance of political matters. But here he fails to mention cultural factors.

2 He passed away in 2015, and many Mongolians suspect that his death might have to do with his fight against mining.

3 I later discovered that Khurts and Baljinnyam have had this office for more than ten years, and the ministry pays a salary for one person. Initially, it was Khurts who received the salary but later this was changed to his colleague, Baljinnyam.

4 In the 1990s, many Mongolians started to bring goods from China and Russia and sold them in Mongolia. From Mongolia, they used to bring marmot skin and scrap metal and sold them in Russia and China.

5 In Chapter 6, we will also see Munkhbayar develop the same argument about individual rights versus the common good.

6 Interestingly enough, not only Khurts but also those who support the liberal economy – namely, Jargalsaikhan – also told me about the same Buddhist understanding of the economy. Jargalsaikhan also said to me that he does not think the market is free and the economy liberal. For this reason, he was uncomfortable being labelled as someone who promotes the liberal economy. Instead, he thinks what he and his other colleagues did in the 1990s – similar to what Ochirbat says in Chapter 1 – was not intended to establish a liberal economy or free market; rather, his sole intention was to assist the economy of Mongolia.

7 Chapter 5 also shows how nationalist movements use these phrases to justify their protests to protect the environment.

8 Although a complete written text was not found, many historians claim that *Ikh Zasag* was an imperial code of conduct of the Mongol Empire and its successor states started from Chinggis Khan.

9 Many others also consider the 1997 Minerals Law to be liberal. For example, Mette High (2012, 255) notes that 'the Minerals Law accentuated the much more liberal position of the state'.

10 'We find ideas about the ownership of land and resources that pivot around the notion of a master-custodian relationship. This model differs from outright individual ownership. Here access to resources, such as pasture, water, etc., is not granted as individual ownership, but rather alternates through an idea of "temporary possession" (*ezemshil*). The notion of temporary possession requires on-going connections that put people in relations of debt to each other, be it between the Mongolian people and its government, between monastic orders and their subjects or between herding households and the spirits or "masters" of local places (*gazaryn ezed*)' (Empson and Webb 2014, 239).

11 He passed away in 2006 due to health reasons. But many Mongolians suspect that his death might have to do with his fight against mining.

12 In 2010, for national security purposes, Mongolian President Elbegdorj Tsakhia introduced a ban on the issue of exploration licences and the assignment of existing permits. He announced that the moratorium was necessary for several reasons: to create better regulation and organisation of mining licences, to amend the minerals law again, to find a resolution to the corruption and errors in the issuance of some thousands of permits and to address the overwhelming sale of land in the form of mining licences (Ninjsemjid 2012). His moratorium lasted for about four years, until 2014.

3
Navigating Nationalist and Statist Initiatives

Mongolians saw a dramatic rise in the mining sector as a result of the Gold Programme and the liberal Minerals Law, as well as other free market and liberal economic principles (see Chapter 1). By the 2000s, the number of gold mining companies and the number of mining licences increased from a dozen to some hundreds. Those who work in the mining companies admit that gold mining companies had enormous importance in the building of the national economy and in funding the nation-state in the 1990s and 2000s (see also the claims of Ochirbat Punsalmaa and others in Chapter 1). Moreover, many Mongolian gold mining companies successfully managed to expand their businesses to non-mining sectors and now present themselves as 'national companies' (*ündesnii kompani*), 'national producers' (*ündesnii üildverlegch*) or 'wealth producers' (*bayalag büteegchid*) that contribute to Mongolia's national economy and development. Hence, they sometimes argue that they can be nationalist in a different sense. More precisely, they are nationalist in the sense of so-called good nationalism (*ündesnii üzel*). In other words, the political appeal to mining and investors invited mining companies to engage in and contribute to the building of the nation and the crafting of the state by helping the national economy (see also Jackson 2015). The appeal also granted mining companies a prestigious reputation at the national level which endorsed their position and influence (for similar cases in Peru and Ecuador see also Bebbington et al. 2008, 2901). In this way, this chapter briefly demonstrates the role of gold mining companies in the nation-building and state-crafting, and shows how they shaped Mongolian politics, the state, the economy, capital and nationalism. In contrast to these politically supported prestigious positions and influences of mining companies, this chapter also shows how pressure from nationalist and statist policies to control natural resources and the resistance of protest movements against mining destruction (see Chapters 4 and 5) forced mining companies to hide, minimise or stop their operations, abandon mining business and start new businesses.

In her book about austerity, Laura Bear (2015) writes about how bureaucrats, entrepreneurs and workers implement and navigate austerity policies. In this chapter, I borrow her use of the term 'navigate' to examine how gold mining companies in Mongolia navigate nationalist and statist initiatives and policies. In the wake of these two forms of initiative, small- and medium-scale gold mining companies faced different challenges, tried different manoeuvres, and indeed many of them reported that they had to reduce or cease their mining businesses and pivot to other sectors. Some used other tactics, such as changing names of their companies or subcontracting other companies to continue their mining businesses.[1] I show in this chapter that as a result of such navigations these companies and their operations became non-transparent. In the words of Marina Welker (2014), mining companies in Mongolia became unstable collective subjects with multiple authors, boundaries and interests. Consequently, it becomes difficult for protestors, authorities and the public to know whether a company stopped its mining operations and businesses, or whether it is simply using a different name or a subcontractor to continue its business. Importantly, this was often *because* of statist and nationalist policies, not *despite* them.

Such nationalist and statist initiatives and policies put small- and medium-scale gold mining companies in a powerless position in terms of defining the nature of their operations. Such an outcome shows something different to what most of the existing literature suggests. There is a conventional framework to understand mining companies and corporations as primarily transnational and powerful institutions. As Stuart Kirsch (2014, 1) puts it, such multinationals 'organize much of the world's labor and capital, shape the material form of the modern world, and are a prime mover of globalization [with] … a wide range of harmful effects'. In this vein, the literature on corporations discusses corporate environmental damage and social disruption versus neoliberal, colonial and capitalist power (Sawyer 2004; Ferguson 2006; Shever 2012; Behrends et al. 2011; Andreucci and Radhuber 2017); sustainability, development and corporate social responsibility (Welker 2014; Idemudia 2010; Raman and Lipschutz 2010; Rajak 2011; Gilberthorpe and Banks 2012; Gardner 2012); audit, accountability and responsibility (Li 2015; Welker 2014); labour and mine workers (Godoy 1985; 206–7,

Smith 2013); and how corporations avoid regulations and respond to critics (Benson and Kirsch 2010; Kirsch 2014).

However, scholars rarely discuss the *absence* of power of small- and medium-scale mining companies in the face of nationalist and statist initiatives such as powerful environmental protest movements, national-oriented political logics and the state control of natural resources. Anthony Bebbington et al. (2008, 2900) point out the relative power of different mining companies to do with their 'sizes and resources' that they 'can use to manage and dissipate conflicts'. Chris Ballard and Glenn Banks (2003, 293–4) acknowledge such problems in the study of mining corporations and note that there is a monolithic image of multinational corporations as 'homogenous, powerful, hierarchical, rational, profit-seeking beasts', which 'tends to mask the considerable complexity of corporations'. In the same ways, this homogenous image also obscures the problems of small- and medium-scale mining companies, investors and Mongolian corporations that I introduce in this chapter. This chapter attempts to deconstruct this homogenous image of mining corpora-tions and show the challenges of some small- and medium-scale gold mining companies in Mongolia. Small- and medium-scale gold mining companies contribute to Mongolia's labour and capital and help to shape modern Mongolia. They are, however, much less powerful than the trans-national corporations that operate in Mongolia, such as Rio Tinto and China Shenhua Energy. I do not focus on the transnational corporations in Mongolia here, not least as the resource conflict in Mongolia started *before* transnational corporations began their operations in Mongolia. In the 1990s and early 2000s, in the absence of the transnational corpora-tions, most of the resource conflicts and resistance focused on Mongolian companies, and later expanded to include some small- and medium-scale foreign mining companies. The absence of transnational corporations in the 1990s suggests that in Mongolia, corporate and community disputes over the environment are not foundationally transnational but have their origin in domestic concerns (see also Hilson and Laing 2016). More precisely, much of the ethnography in this chapter concerns Mongolian gold mining companies with no foreign investment or shareholder in the ownership of the company. Only the last section of the chapter presents a medium-scale foreign gold mining company and its struggles to navigate those nationalist and statist initiatives.

In addition to the problem of the homogenous image and assumed transnationalisation of mining companies, there is a lack of a proper account of mining corporations and companies. In their literature review, Ballard and Banks (2003, 290) address the lack of attention to corporations and criticised anthropologists' preference to focus on the 'exotic', centring on the position of local communities in the vicinity of multinational corporations. A decade later, Emma Gilberthorpe and Dinah Rajak (2016, 6) make the same statement. They claim that the contribution of anthropology to the study of natural resources should be to bring into focus the agency of corporations including the smaller companies that I describe in this chapter. Kirsch (2014, 12; see also Benson and Kirsch 2010, 459), who worked with protest movements, reveals the same methodological and ethical concern and admits that it was impossible for him to adopt a middle position in the conflict between protest movements and mining companies as I mentioned in the preface. However, I choose to take a multi-faceted approach, in order to achieve an accurate balance in my presentation of conflicts between the community and company. This approach permits me to present different agents such as activists, companies and politicians in the same account and shed light on all the parties involved. In this way, my focus on both gold mining companies and those working in the mining sector is intended to make a contribution to understanding corporations and companies from a more balanced and comprehensive perspective.

While Chapter 1 described the establishment of the liberal mining economy and the reification of the national economy, and Chapter 2 examined the growth of national-oriented legislation and resource control – and the following chapters will focus on protest movements – here it is necessary to discuss what happened to the gold mining companies. In other words, this chapter concentrates on one of the other sides of the story; it depicts the experiences, positions and justifications of gold mining companies. There are many questions that often remain unanswered: What happened to those companies? Why did they cease operations? What happened to them when they ceased? What manoeuvres did they use to endure? I collectively refer to these under-studied and unanswered questions as the 'other side of the story'.

This chapter shows the different navigations of three gold mining companies, which helped them to manoeuvre, manage and overcome the challenges of some nationalist and statist initiatives. In the first case of Erel, to combat environmental protest movements the company founders and owners attempted to use their political networks and influence by establishing a political party and control of the government in coalition with the Democratic Party. In the second case, the Gatsuurt company, its founder and owner L. Chinbat, although he criticises it, acknowledges the power of the culturally sanctioned nationalistic and statist initiatives, and considers the precariousness of the resource business in Mongolia.

The above-mentioned companies also show how capital generated from mining flows to other sectors by establishing other companies at that buy other resources in those sectors. Unlike Erel and Gatsuurt, Cold Gold Mongolia, a New Zealand company, continues its mining business. However, to continue its mining business the company had to execute different manoeuvres to navigate varied nationalist and statist pressures using its Mongolian family network. A description of these three companies' 'navigations' reveals the manner in which mid-range mining companies were forced to grapple with Mongolian national politics, which in turn arose from a very particular political-economic and, indeed, cultural context.

Erel Company: The Motherland Party

The Erel company plays an important role within this book. Not least as the 'river movements' that I will present in the following chapters started their protest against the operation of Erel and other, smaller companies along the river Ongi. During my fieldwork in spring of 2017, my attempt to interview the director of Erel and other people from the company failed and it was very difficult to get firsthand information about this company. Therefore, I briefly present Erel based on the available materials from the media and some interviews with environmental protestors and people at the Mongolian National Mining Association. As the company describes on its webpage, although it started its business from geology and mining in 1989, from 1994 it expanded to include construction, education, banking, finance, management and real estate. There is no mention of mining,[2] which is the first example of an act of 'navigation' to escape from its mining destruction scandal by removing that practice from its own history. For those who suspect that Erel still profits from mining, such a navigation leads to the creation of an opaque state of affairs.

Not long after the collapse of socialism, when private property and the free market were still new in Mongolia – that is, in the 1990s and 2000s – Erel was one of the first and largest privately-owned mining companies that started its growth from open-pit gold mining. According to Bayarsaikhan Namsrai, one of the leaders of the river movements, at the time when they started fighting against Erel, the company had already managed to establish 13 other businesses in addition to its mines, including a bank, a construction company, a secondary school, a hospital, an insurance company and more. They were also the most significant taxpayers in the country. In other words, the protestors had to fight against the biggest so-called *bayalag büteegch* (wealth producer) in Mongolia of that time. Alongside his 13 businesses, in 1998 the owner, B. Erdenebat, founded the Mongolian New Democratic Socialist Party (*Mongolyn shine ardchilsan sotsialist nam*) and not long after changed its name to the *Ekh oron nam* which literally means 'Mother Land or Mother Country Party' (see also High 2017) to mobilise the culturally salient nationalistic tendencies surrounding Mongolian concepts of land and territory discussed in the previous chapter. I argue that the change in the name of his political party is a form of navigation to re-introduce the image of his mining company to Mongolians. The nationalist presentation of the party helped him to succeed in politics by appealing to ideas and discursive resources that were attractive to many citizens. Erel was an extremely profitable company and its political party was initially successful. Erdenebat managed to establish his political party offices in almost all provinces and districts across the country. In 2000, in the parliamentary election, 73 candidates ran for the Motherland Party, but only the party chair and the company owner, B. Erdenebat, were elected to parliament. In 2004, in the next election, the Motherland Party won seven seats in parliament, which was a tremendous victory, and formed a coalition with the Democratic Party. In the new government, Erdenebat worked first as a Defence Minister (2004–5) and then as a Fuel and Electricity Minister (2006–7). Other company men such as I. Erdenebaatar, former CEO of Erel, served as a Minister of Environment, and S. Otgonbayar, also from Erel company, a Minister of Emergency.

However, the success of Erel and the Motherland Party in Mongolian politics did not last long. The coalition collapsed: according to the media, because Erdenebat (the chair of the Motherland Party) and Enkhsaikhan Mendsaikhan (the chair of the Democratic Party) fought over the prime minister's position (Mongol News 2011a). The Motherland Party started to receive heavy criticism regarding the environmental destruction inflicted by its related companies. At the same time, as the founder and leader of river movements Munkhbayar Tsetsegee claims, the river movements strongly pressured (*dömögkhön shakhaj baisan*) Erel by collaborating with some of the politicians fighting against Erel for different reasons (see Chapter 4). For those politicians fighting Erel, Munkhbayar and the river movements became an active ally to defeat the company. Mette High writes the following about the situation:

> Munkhbayar was relentless in his efforts, and eventually, parliamentary commissions conducted the promised research in 2007. Followed closely by the public, the inquiry investigated

several mining sites, including Erel's operation in Ölt. In the interview published in a widely circulated Mongolian newspaper, a member of the commission laid bare the conclusion that Erel had been purposely forging the environmental rehabilitation procedures mandated by Mongolian law. He stated that 'the responsibility for all the mess created in Uyanga *sum* should rest with Erel Company' (Ödriin Sonin 2007). Several members of parliament began to attack Erel for the situation in Uyanga openly. Then a debate followed, broadcast on the state funded TV network MNB (Mongolian National Broadcasting). The owners of Erel declined to participate, and in their absence, the debate placed the responsibility for the 'ecological catastrophe', as the State Property Committee chairman Zandaahüügiin Enhbold described it, squarely on Erel's shoulders. . . . Declaring Erel, the sole perpetrator was not only a potentially fruitful legal strategy but also an attempt among politicians to wage their own political battles. (High 2017, 56)

In the next election in 2008, the Motherland Party did not win any seats. There are rumours that Erel sold its mine to a Czech company called Aum Alt (Aum Gold). However, there is no information about a company called Aum Alt or Aum Gold in either English or Mongolian. Munkhbayar suspects that Erel has not left mining completely. Instead, he thinks, the company became hidden by having multiple layers (*dald orchikhdog*) of other companies up front, while Erel remains in the core. Munkhbayar claims that they pretended to sell their licences to other companies and then subcontracted them to perform much the same activity (see also High 2017, 57).[3] In the case of Erel, it is difficult to determine whether Munkhbayar's suspicion has some truth or not. Therefore, it is difficult to claim that there is hidden control or show how this works in the case of Erel. Yet I do not deny the possibility of other companies that can work through subcontracts to help navigate challenges.

Erel's political career did not end up helping the company; instead, it brought a decline to its business. The political power of the company did not help it solve its conflicts with environmental protesters. Indeed, the situation turned out more or less the opposite. Erel's involvement in politics helped the river movements to bring the issue of mining-induced destruction and their nationalist initiatives to a higher level of political relevance and to receive recognition and support from politicians that fought against Erel and the Motherland Party. It was also a bad financial move for Erel to establish a political party with hundreds of offices across the country. According to Algaa Namgar, it can be assumed that Erel spent too much money to fund the Motherland Party, and as a result, the operations of its offices across the country and the election campaigns of its candidates severely damaged the company's financial security. Despite its failures in politics, the success story of Erel and the Motherland Party remains an excellent example of a gold mining company in the contest of the nation-building and state-crafting in Mongolia (see also Jackson 2015).

Besides the decline of its power in politics, at around the same time in 2007, the Erel mine faced strong resistance from local protestors. Munkhbayar and his colleagues stopped the operation of 35 gold mines out of a total of 37. Erel had to stop its mining operation for about three months. Furthermore, in addition to the effect of the protests of river movements, the 2006 windfall tax and the 2009 'law with the long name' (Law to prohibit mineral exploration and mining operations at headwaters of rivers, protected zones of the water reservoir and forest area)[4] had a major impact on gold mining. I turn now to an ethnography of two companies, which show how some of the gold mining companies survived such nationalist and statist initiatives.

Gatsuurt Company: The National Food Producer

Chinbat Lhagva, founder and owner of Gatsuurt Company, claims that all of its mining operations ceased[5] due to the 68 per cent windfall tax and the seizure of all mining licences in accordance with the 'law with the long name'. For this reason, the company is currently known by the Mongolian public not as a mining company but as one involved in agriculture and food production.

Founded in 1992, Gatsuurt is one of the largest, wealthiest and most successful Mongolian companies. Chinbat was born and raised in a herding family of 15 children. As is the case for many environmental protestors, Chinbat, his parents and many of his relatives were herders. Professionally trained as a geologist in the Soviet Union, after graduating in 1985 he started a job at the International Geological Expedition (*Olon Ulsyn Geologiin Ekspeditsi*). In 1991, he had an opportunity to study satellite geological exploration for one year in France. When he returned from France, the expedition was closed. To make his living he was involved in *naimaa* (suitcase-trade) between Beijing, Ulaanbaatar and Moscow for a couple of years (for *naimaa* see also Ichinkhorloo 2018). His experience working in geological exploration and the profits made from trade allowed him to establish a mining company, Gatsuurt, which started gold mining exploitation in 1995 and 1996. The company negotiated a bank loan of 200 million MNT to buy some old Russian mining equipment in order to start mining operations. Not long after, he experienced the disadvantages that come with old equipment and technology. About the same time, he met Danny Walker, a miner from New Zealand. What he learned from Walker made him decide to visit New Zealand to learn more about mining equipment and advanced extraction and rehabilitation technologies. After he returned, Chinbat got another bank loan to renovate his equipment and technology, this time for US$1 million from Anod Bank of Mongolia (see also Batsukh and Chinzorig 2012). He explained to me that mining extraction technology in the 1990s was very backward (*ashiglaltyn arga technologi mash khotsrogdson baisan*). Companies used to dig massive holes and left them for years, rather than fill up the extracted land immediately while continuing to extract more land.

With Soviet mining technology, all the material gets extracted and then it is all filled up later. With the modern mining technology, filling up and extracting go together. According to Chinbat, protests of local people started when companies left the extracted land unfilled and without rehabilitation. In the countryside, most Mongolians who consider digging up the ground a taboo (Shimamura 2014, 395) felt uneasy with the extracted surface of the ground. As someone who grew up in herding culture, Chinbat understands the potential of this residue and degraded surfaces of the ground to trigger emotional reactions from many Mongolians. The Soviet technology was also financially much more expensive than the New Zealand technology. In the late 1990s, Gatsuurt started to utilise a new technology that Chinbat gained from New Zealand, and immediately started both technical and biological rehabilitations in the late 1990s (see also Batsukh and Chinzorig 2012). Unlike Erel, the technology helped him to avoid being targeted by environmental activists for protest against mining-induced environment damage. In 2001, the Ministry of Environment approved its first technically rehabilitated 3.5 hectares of land, and both technically and biologically rehabilitated 24.8 hectares of land in 2002. The size of the rehabilitated land increased from year to year from 3.5 hectares in 2001 to 268.4 hectares in 2010. Also, in 2001, the Ministry of Environment awarded Gatsuurt a certificate of 'Number One Rehabilitation Company of Mongolia' (*Mongol ulsyn neg nomeriin nökhön sergeegch company*). At that time, only a few companies were conducting environmental rehabilitation. Therefore, in 2001 in Bat-Ülzii *sum*, Övörkhangai *aimag*, Gatsuurt organised a national workshop on mining rehabilitation (*ulsyn khemjeenii nökhön sergeeltiin zövlögöön*).[6]

Around the same time, Gatsuurt started interacting with environmental protestors, including Munkhbayar. With the help of the above-mentioned new technology and lawful operation, Gatsuurt did not get into a severe fight and conflict with the environmental protestors. However, as Chinbat admits, this is beside the point. Even though there was no direct conflict between Gatsuurt and the movements, and environmental movements did not target Gatsuurt, the company was left financially harmed because of the overall impacts of the later environmentalist and nationalist initiatives of the windfall tax and the 'law with the long name'. In consequence, Chinbat had to stop his gold mining operations.

In 2006, after the approval of the windfall profit tax law, I heard Chinbat say that the law made it impossible to engage in gold mining in Mongolia. He complained that there was no way to make a profit after paying the 68 per cent windfall tax. He later explained to me that the high tax left no profit for gold mining companies. For this reason, for example, Gatsuurt had to speed up the exploitation of the most significant and least costly deposit in two years. He was forced to leave the rest of the minor deposits untouched due to the high cost of extraction and high tax to sell. He further explained that incomplete mining extractions that left some gold behind helped to expand the extraction of illegal artisanal mining, which is much more harmful to the environment (for artisanal mining see High 2017). Consequently, the company sold the gold extracted from the most efficient deposit in 2007 and 2008, when the price rate was low in comparison to the increase in the international market in the following years, which reached as high as US$1,700 in 2013. For Chinbat and many other gold mining companies, it was suboptimal to extract and sell the main deposit in such a short period and sell them at a low-price rate. He complains that the success of Mongolia's gold mining economy and hopes of gold mining companies was tarnished and ultimately terminated by such nationalist and statist initiatives.

By 2009, in the aftermath of the windfall tax, the situation for gold mining companies worsened with the approval of the 'Law to prohibit mineral exploration and mining operations at headwaters of rivers, protected zones of the water reservoir and forest area' or the 'law with the long name', which was drafted and lobbied by 'river movements' (see also Chapters 4 and 5). As a result of the law, about 20 gold mining extraction licences of Gatsuurt (see also Sansar 2015) were voided or stalled due to the new legislation. This time, the legislation made it impossible for Gatsuurt to operate at all. Chinbat explained that the state 'snatched' (*tör khuraagaad avchikhsan*) all of his licences. He now owns a small per cent of a share in a small gold mining company and most of his income comes from non-mining businesses. When I asked him if he will return to the gold mining business he indicated that he wouldn't, due to the amount of work and investment it would take to get the necessary permissions, licences and equipment. Furthermore, he complained that in Mongolia the state can always grab mining licences which makes the business precarious. In other words, Chinbat emphasises the impact of nationalist and statist initiatives and policies which have gained salience in the political logics of Mongolia.

According to Chinbat, if the windfall tax and the 'law with the long name' had never happened, his company and many other gold mining companies would have made fortunes, and the Mongolian economy would have benefitted enormously from the increase in the price of gold in the international market. He could have sold the gold deposits throughout the periods of the price increases with much higher rates up until 2013. According to his calculations, Mongolia would have increased its gold production from 43 tonnes in 2006 to 100 tonnes in 2011. He claimed that Mongolia lost around US$50 billion. He also speculates that about 400 tonnes of gold (about US$14 billion) were smuggled into China.

After all of these struggles in the mining sector, he concluded that the most secure business was food production, which is not as precarious as mining and does not receive the same nationalist and statist pressures. The company's shift from mining to agriculture and food production shows how capital works in Mongolia. In a similar way to Erel, the company was already investing its profits from gold into other businesses such as agriculture, food production, five-star hotels and television. Although the company started with gold mining in the mid-1990s, agriculture and food production were not entirely new for the company. Gatsuurt began to operate in these fields in the early 2000s. From the mid-2000s, Gatsuurt successfully diverted its primary focus and the overall identity of the company away from mining to agriculture and food production. Consequently, Gatsuurt now supplies about 15 per cent of wheat, 5 per cent of vegetables and 2 per cent of meat to the domestic market of Mongolia.[7] In this way, Gatsuurt positioned itself as a national food producer not a mining company, and managed to build a better reputation than many other mining companies. In 2017, the president of Mongolia granted Chinbat the highest and most prestigious distinction of the country: 'Hero of Labour' (*Khödölmöriin baatar*). Considering the often negative reputation of mining, many in Mongolia assume that the distinction celebrates his achievements in the agriculture and food production, rather than in mining. I assume that the decline of his mining career and his achievements in Mongolia's agriculture and food production sector granted him an opportunity to receive this distinction. In other words, it would have been difficult or maybe impossible for Chinbat to become the 'Hero of Labour' if he was solely a mining company owner. His successful shift from mining to food production and other businesses in the country allowed him to receive the award.

Besides helping the corporate image and reputation, the diversification of the businesses of gold mining companies in the case of Erel and Gatsuurt, and their shift of capital in critical times functions as a form of navigation to minimise loss of potential risks in the precarious mining business in Mongolia. However, those with no diversification opportunities, or those not prepared to make a shift in their capital, must find alternate forms of navigation to stay in the mining business, as we will see in the case below.

Cold Gold Mongolia (CGM): A Solution to Mining Destruction

Cold Gold Mongolia Co. Ltd. (CGM) is an open-pit gold mining company established by Danny Walker, a miner from New Zealand. Unlike Erel and Gatsuurt, CGM still runs a business in the gold mining industry in Mongolia. However, Danny told me that CGM no longer directly operates mining in Mongolia, but only manages mining operations for Mongolian gold mining companies. With this statement, he meant CGM no longer holds licences for mines that CGM manages. More precisely, as Danny complained, different pressures made it impossible for a foreign company to hold a mining licence. The only way left for CGM was to manage mining operations for Mongolian mining companies that hold licences. Such a strategy insulates CGM from different pressures, which is a further form of navigating nationalist and statist initiatives.

Danny presented his company as a model of a solution to the environmental damage caused by mining in Mongolia. In contrast to Erel and Gatsuurt, CGM has a different corporate image based on its rehabilitation technology. Using the advantage of its mining technology, CGM usually operates in the sites previously mined with Soviet technology and terminates mining projects through rehabilitation. By properly terminating mining projects CGM prevents mined areas from further degradation caused by illegal artisanal mining (see also High 2017). Without the company, many mined sites have been left for many years without rehabilitation. Therefore, CGM presents itself as a company with the solution to mining-induced environmental damage which helps the company to justify its operation.

In 1994, the Resource Management Amendment Act in New Zealand (New Zealand Parliament 1994, Article 30) endowed local authorities with the right to regulate permissions to use water and disturb the surface of the ground. The amendment made mining difficult, and many mining companies fled the country. One of them was Danny Walker. He mainly operated underwater gold extraction in New Zealand and had a company called Under Water Mining. In 1998, while working in Indonesia, he heard about Mongolia from a Canadian miner friend who worked in Mongolia at that time. After visiting Mongolia in 1998, Walker started Cold Gold Mongolia open-pit gold mining company in 1999. According to Walker, he decided to work in Mongolia for two main reasons. First, the overall social, economic and political environment to operate and run a mining company in Mongolia was welcoming and friendly enough, compared to many places he observed and experienced. Second, Mongolia was attractive with its rich and large gold deposits, where a company could base itself for some years in order to grow. According to Walker, it was not only him and his company but also many other foreign companies that were very much attracted to Mongolia, including AGR Limited, owned by Resolute Australian mining company, which started its operation in the Boroo Gold Mine in 1999. In early 2002, Cameco (Canadian Mining and Energy Corporation), later known as Centerra Gold, entered Boroo Gold Mine. Walker was the first to introduce the New Zealand method of placer gold mining of stream beds in Mongolia, the one which Chinbat learned and adapted in this company. The method uses a 'mobile screening plant' and 'hydro-active riffled sluices'. This technology 'permitted high percentage gold recovery, 24 hours a day gold washing, same day rehabilitation of mined out areas, reduction of truck movements and efficient use of manpower' (Walker 2001, 36). This method was revolutionary compared to the Soviet method widely used by most of the Mongolian and Russian companies in Mongolia. In 1999, CGM started its operation in the abandoned workings of the Toson Terrace in the Zaamar Goldfield, which was previously mined by a Mongolian company from 1993 to 1996 using traditional Soviet mining and washing methods. They produced 460 kilos of gold from 1993 to 1996, and their gold recovery was 51 per cent (Beaudoin 2000). In a four-month-long scavenging operation from previously extracted workings, CGM produced 37 kilos of gold (Walker 2001, 36). At the same time, CGM was one of the first – and only – gold mining companies that started rehabilitating mining areas. In 2001, the company received a distinction from Selenge province (*aimag*) government where the company mined: 'Environment Friendly Organization'. After Selenge, in 2004 CGM moved to Urtyn Gol mine in Shine Jinst *sum* (sub-province) of Bayankhongor *aimag* (province). Gatsuurt company had previously mined Urtyn Gol deposit and sold it to CGM.

In Urtyn Gol, in 2005, CGM first encountered local resistance. First, it was one or two people coming to the mine who expressed their resistance to mining, despite everything being legal and following all standards and regulations, as Walker claims. The local herders' resistance against CGM started to enlarge and became more active in 2006. Local herding families in the area joined and arrived with a Russian van, and some 20 people came to protest against the mine. They demanded that the company stop its extraction. Their primary concern was to not let the mine extract areas that were covered with broom-grass (*ders*), and to preserve those areas for pasture. The company had to negotiate with local herders and decided to mine in an area with less or no broom-grass. However, the areas that local herders permitted CGM to mine did not necessarily have good gold deposits.

At that time, Walker married Orgilmaa Zundui-Yondon, who was the office manager of CGM. She told me how frustrating it was for a foreign man and Mongolian woman to deal with the resistance of the local people. Because of such difficulties, the company stopped its operations the following year, rehabilitated the area and terminated the project. CGM fenced the mined area, used seeds to plant grass and watered the rehabilitation area. Some local people, such as Erdene, participated in the rehabilitation. Erdene was one of the local herders who actively resisted and led the demonstration. Walker and his wife came back to see the rehabilitation three times in the following years and became good friends with Erdene. In 2006, the General Agency for the Specialised Investigation of Mongolia started awarding a distinction to mining companies that successfully conducted rehabilitation (*nökhön sergeeltiin tergüünii company*). With the rehabilitation in Urtyn Gol, CGM was one of the first four companies to receive this distinction.

CGM's next mine site was in Daldyn Am, also in Bayankhongor, in a different *sum* called Bömbögör. This time they managed a mining operation for a Mongolian company called Khan Shijir. Before CGM's operation, another Mongolian company called Erkhis Mining operated the Daldyn Am mine for Khan Shijir. Erkhis Mining stopped its operation for Khan Shijir due to the 68 per cent windfall tax. As Orgilmaa explains, according to the calculation of CGM, the Daldyn Am mine gold deposit was large enough to make a profit despite the increased windfall tax burden. Considering this calculation and CGM's desperate situation to start a new mining operation, the company came to an agreement with Khan Shijir, and CGM started operating in the Daldyn Am in 2006 and 2007. Here, Walker and Orgilmaa experienced the difference between managing a mining operation and holding a mining licence. As an operation manager, and not a mining licence holder, Walker and Orgilmaa were both kept away and secure from local resistance and harassment. Since 2006 local resistance against mining and environmental movements in the whole country had become more influential than ever before (see Chapters 4 and 5). However, Daldyn Am mine faced relatively less resistance and struggles because, as Orgilmaa explains, the mining licence owner was a Mongolian company and those interacting with the local people on behalf of the company were Mongolian men of the company, not a foreign man married to a Mongolian woman. The owner of Khan Shijir, Dovjid, has two grown sons, and both of them had been working and residing at the mine site. The two sons used to deal with the local herders' concerns and disputes. Orgilmaa said that the people in the movement and local herders in Daldyn Am treated Khan Shijir much better compared to her experience in Urtyn Gol. From their experience working for Khan Shijir, Walker and Orgilmaa learned that Mongolian companies led by Mongolian men would face fewer problems, compared to foreign companies. With the help of the Mongolian mining company dealing with conflict and resistance, CGM successfully closed the mine with excellent environmental rehabilitation. Consequently, in the following year, Khan Shijir won the distinction of the best rehabilitator of the year from the General Agency for Specialised Investigation of Mongolia.

After the Daldyn Am mine, CGM started operating a mine in Nariin Teel, in Bayankhongor. About 100 people from three *sums* along the Nariin Teel river protested against the mining operation, and the company had to move again to a different mine in Khüitnii Am, which was in Bömbögör, Bayankhongor, and started operating in 2008.

In an interview in 2017, Walker told me that 'no one [foreigners] makes the bureaucracy, no one likes us here'. Because of this, Orgilmaa established a Mongolian company called Odod Gold for the next mining project, using the facilities and equipment of CGM. The establishment of a Mongolian company was a necessary form of navigation to deal with local resistance against mining and bureaucracy. The Khüitnii Am mine was a long and narrow gold deposit that occupied pastures of about 20 herder families. While the company already had all of the necessary regulations and permits, those 20 families were the first to fight against the mining operation. The first among these protestors was an elderly lady whose pasture was the first to be exploited. She brought a gun with her and threatened to shoot anyone who came to extract from her land. She was not the only person who brought a gun; many others did. The company ended up negotiating with the families and agreed to pay one million MNT to operate one screen gold wash plant per month. The company had nine screens and had to pay nine million per month. Orgilmaa says it was an excellent financial contribution to those families. Some of them bought more livestock, some bought houses in the provincial town, and some paid student fees. Also, Orgilmaa told me, no one and no regulations helps mining companies to stop local resistances. More precisely, there is no formal arrangement or assistance from the central or local governments or police when local people resist mining even when mining companies have all the necessary legal paperwork and permits. Instead, the state often leaves such incidents for the mining company and local people to decide amongst themselves. Within the range of their freedom to negotiate, Orgilmaa and her family paid local people to get their permissions. However, the real difficulty comes when suddenly a new law stops mining operations.

As mentioned in the Introduction, on 16 July 2009, the 'Law to prohibit mineral exploration and mining operations at headwaters of rivers, protected zones of the water reservoir and forest area' – prohibiting mining operations within 200 metres of river and forest areas – passed, voiding many mining licences and forcing a large number of gold mining companies to terminate their on-going operations and future plans. For example, Odod Gold had to halt their activity in Khüitnii Am mine, and had their licences suspended for future operations. Odod Gold and many other companies were not prepared for this. In these extreme conditions, those who were willing to continue gold mining had to buy new mining licences in legal areas – that is, away from rivers, lakes and forests, which was not typical for open-pit gold mine deposits.

It took almost three years for the government to identify and calculate the size of target areas, impacted companies and licences, and to figure out how to implement the law, and most importantly, how to compensate[8] those mining companies that stopped their operations. As Walker and Orgilmaa understood the situation, in 2010, neither parliament nor government required those companies to stop their operations immediately. Instead, the government approved annual work plans for gold mining companies as the 'Law to prohibit mineral exploration and mining operations at headwaters of rivers, protected zones of the water reservoir and forest area' was violated. The unclear situation left a severe and insoluble conflict between movements and companies. Odod Gold – and many other companies – had their annual work plans approved and permitted by the government. However, protestors considered mining operation in such circumstances a violation of the law and pressured mining companies and the government to implement and follow the law. Odod Gold did not immediately stop its operation: they operated in Khüitnii Am for only one year, which was not a long enough period to create new plans and facilities for their next operation, and to replace all their suspended licences. Further, to cease its operation in Khüitnii Am, such a short extraction time was too great a financial loss for the company. As a result, Odod Gold continued to operate and struggled through chaotic conditions throughout 2009 and 2010.

Bömbögör *sum*, with a population of about 3,000, had four different protest movements against mining. These were in addition to the bigger movements established in the provincial centre, such as *Khongor Nutgiin Duudlaga* (Call of the Khongor Homeland), one of the 11 movements that joined the 2005 *Mongol Nutag Minu Evsel* (My Mongol Homeland known as Mongolian Nature Protection Coalition) united by Munkhbayar (see Chapter 4). Following the new law and throughout 2010, almost every week people from different movements visited the mine in Khüitnii Am, travelling from *sum*, *aimag* and the capital city to demand that operations be stopped. The whole situation persuaded those 20 herding families at the Khüitnii Am mine site to strengthen their resistance against Odod Gold. The only way to negotiate was to increase the monthly payments to those families.

In 2011, the government stopped giving gold mining companies permission to operate and provided no compensation. Many companies continued to operate illegally without state approval or permission. Others declared bankruptcy, while many others searched for alternative business, as was the case for Erel and Gatsuurt. Those who illegally operated had to spend fortunes to pay local people, movements and authorities. In the same way, Odod Gold operated illegally in order to avoid bankruptcy and to generate funding for a new mining operation in legal areas. In this way, the whole situation – that is, with no compensation from the government – eliminated many companies and put them into desperate positions which in turn led them to operate illegally. Walker and Orgilmaa remember this as their toughest year, and only those companies that managed to secure funding for their next mining project survived. The consequences of their illegality led them to experience all sorts of pressure from the national and local government authorities, local herders, movements, artisanal miners and even their employees. For Odod Gold, being a successful mining company with the record of highest possible gold recovery, advanced operation technologies, minimal environmental destructions, best practices of rehabilitation, fair tax payment, contributions for rural development and national reserves of gold did not help them to survive the year. It had been Mongolia's policy from the 1990s to encourage and enlarge the mining sector and appeal to many mining companies, so of course, many in the mining sector considered that it was unfair to sacrifice mining companies, especially those that had previously strictly followed the law.

Ultimately, as a result of the turmoil, which lasted for about three years, Odod Gold finally stopped its operation in the Khüitnii Am mine. They conducted rehabilitation in the mined areas and planted about 18,000 trees and bushes, and then abandoned their remaining gold deposits in the mining field. After struggling for three years, Walker and Orgilmaa finally managed to start a different mining project. In 2012, Danny and Orgilmaa moved away from Bayankhongor to mine in Zaamar, Töv *aimag*. As a result of the president's moratorium, it was indefinitely forbidden to sell mining licences and to explore new gold mine sites. Because of these constraints, CGM and Odod Gold had come up with new techniques of navigation, and this time they had to collaborate with a different mining company with licences to operate in the permitted areas. Z. Batbaatar, Orgilmaa's brother, owns a mining company called Uuls Zaamar. When I visited Danny and Orgilmaa at the Uuls Zaamar mine in spring 2017, CGM, as Danny explained to me, was managing a mining operation for Uuls Zaamar. They were about to finish environmental rehabilitation of the mined area and had started operating the neighbouring site left by the Russian company Golden East Mongolia (Altan Dornod Mongol), which now has a Mongolian owner.

Zaamar is a site with a long history of gold mining dating back to Soviet times. This history makes mining in Zaamar different from Bayankhongor. In 2012, when Uuls Zaamar started its operation, the company faced a different conflict with the residents. This time the residents were artisanal miners, not herders. Those miners live in a small settlement only a few kilometres away from the Uuls Zaamar mine site. This tiny settlement is often referred to as *güür* (the bridge), indicating the close location of the settlement to a bridge. Within 15 kilometres, there is another much larger mining settlement called Sov, also populated mainly by artisanal miners, possibly dating back to the Soviet period. One evening not long after Walker and Orgilmaa settled at the Uuls Zaamar mine site, local artisanal miners protested against the company and demanded to be granted areas to operate in. This demand was legal due to the regulation lobbied for by the Swiss Agency for Development and Corporation in Mongolia to support local artisanal miners forming small cooperatives (*nökhörlöl*). Orgilmaa, together with two of her brothers and members of the company, had a meeting with the representatives of the protestors and explained why its mining technology and the deposits' location deep under the ground was not suitable for artisanal miners to exploit. The protestors nevertheless insisted that the company grant them immediate access, and warned them that the protestors gathered outside the mine were ready for violence (*yu ch khiij magadgüi*). Nevertheless, the company made a decision not to let the artisanal miners work and warned them that the protestors would be charged for their illegal and violent acts. The company employed around ten security guards to protect the mine site, and they were prepared to fight against the strike of local artisanal miners. Orgilmaa said that there is always some resistance when a mining company starts operation and the company should be prepared for this. The beginning is always a crucial moment, which decides what will happen to the company in the next couple of years. The only place they could not negotiate at all was in Nariin Teel in Bayankhongor, where the company did not operate at all and left (see above). This experience caused Orgilmaa and her family to make the tough decision in Zaamar: to fight against the protests of the artisanal miners, instead of accepting protestors' requests and trying to negotiate. Mining company's aggressive responses to the resistance of protest movements is another form of navigation, perhaps the fiercest and the most desperate kind, compared to earlier examples we have seen in this chapter.

To conclude, the ethnographic material in this chapter allows one to understand agency within the management of small- and medium-scale gold mining companies. To understand small- and medium-scale gold mining companies and their operations in Mongolia, one has to focus on the individual agent or agents who own and manage the company or companies in collaboration (for the same situation of other small and medium businesses in Mongolia see also Chuluunbat and Empson 2018). Although as a mining company Batbaatar owns Uuls Zaamar, and *this* is the company that officially holds the mining licence, family members who own other companies share different duties and operations of Uuls Zaamar. Walker explains that it is now his family who mines in Zaamar;

his company, CGM, only manages the mining operation. The relationships and interconnections of these companies, which are owned by different family members, appear to me an entangled relationship of three different mining companies, and the complexity of their relationship creates some level of non-transparency which may be the key to survival. What is clear is that the arrangements made between these three companies enable Walker, Orgilmaa and Batbaatar to manage or operate mining and continue their business in the open-pit gold mining sector. Considering such a complex relationship, one cannot discuss a monolithic or even singular company. Instead one must trace the agents who own these companies and manage such tangled relationships. Walker and Orgilmaa's story shows how they managed to survive in Mongolia's mining sector. The entanglement of these three companies is a by-product of Mongolia's uneven gold mining policy in the past two decades. The law and regulations made it impossible to have one company sustainably operate in a transparent manner – instead, the collaboration of different companies is required. Furthermore, the presidential moratorium temporarily banned any form of exchange or sale of mining licences and exploration of new mines, so the only way to access and use a new mining licence was to buy or to collaborate with a company that possesses such a licence. In this way, some companies closed down their operations while others hid within the entangled relationships between multiple mining companies. Such relationships, along with a general lack of transparency, protect companies or their owners from possible problems and conflicts. In Mongolia, a mining company and its operation is a multiple and complex process of navigation of sometimes desperate individual agents acting on behalf of different companies in the unpromising environment of the resource economy brought about by *national* as much as *international* political-economic processes.

Notes

1 I must make clear that I do not argue that environmentalist and nationalist movements and state control are the only causes of the decline of these gold mining companies. There are many other related factors and causes to do with their technology, management, accountability, expertise and investment and the fall of price in the international market as some environmentalists, politicians, gold mining companies and experts in the mining associations note and acknowledge.
2 For more information see http://erel.mn.
3 Chapter 4 and 5 show how the movement successfully closed down many mining operations and stalled mining licences. These chapters suggest that local river movements are powerless in the 'mining war', but they are a powerful group in the battle.

4 As I mentioned in the Introduction, the official name of the law is *Gol, mörnii ursats büreldekh, usny san bükhii gazryn khamgaalaltyn büs, oin san bükhii gazart ashigt maltmal khaikh, ashiglakhyg khoriglokh tukhai khuuli* (Law to prohibit mineral exploration and mining operations at headwaters of rivers, protected zones of the water reservoir and forest area). Munkhbayar, the leader of the river movement, complains that nicknaming the law such as the 'law with the long name' is a political tactic to obscure the indication of the law. For him, the intention of the law that appears in the 'long name' of the law gives a proper indication (see Chapter 5).

5 Many other companies also reveal the same. For example, the historical outline of the Altan Dornod Mongol (Golden East Mongolia; est. 1997) gold mining company, on its webpage, says that the company did not operate from December 2009 to September 2011 due to the National Police Agency investigation. Furthermore, the website states that due to the stall, the company owes 360 billion MNT, 1,000 hectares of the field was left without rehabilitation and caused enormous environmental damage, and all the equipment was left unattended and was therefore looted and eroded. For more information, see www.adm.mn/mn/бидний_тухай/с/Түүхэн_замнал.html

6 For more information, see the company website www.gatsuurt.mn

7 For more information, see the company website www.gatsuurt.mn

8 Later official calculations of the government revealed that the government had to compensate about 222.5 billion MNT which was two or three years' GDP (Bilguun 2015).

4

Advocacy and Activism in Popular Mobilisations

This chapter provides an account of a series of mobilisations that began in 2001 with the Ongi River Movement (ORM) to protect the Ongi River from mining-induced environmental damage in the south of Mongolia. In reference to these movements, Dalaibuyan Byambajav writes that:

> The interaction between local, national, and transnational actors constituted dynamic and contested framing processes. The discrepancy between local values and opportunities and transnational norms had a significant effect on the trajectory of the movement. The models of local movements that the foreign donors sought to promote in Mongolia mutated into different 'blended forms'. Most local movement organisations are increasingly transforming into advocacy organisations rather than community-based organisations that the donors anticipated (Byambajav 2015, 97).

His observation of the difference between advocacy organisations and community-based organisations is something that interests me in this chapter. Although there is much overlap and connection, the ethnography of this and the next chapter allows me to take the difference further and show how the ORM did not have its origin in the activism of ordinary local people and herders. Rather, it was originally a working group of local, regional and national officials, elites and experts with important positions in the local and central governments, ministry, academic institutions and national media, and received support from different international donors with the capacity to advocate through different channels. I use the term *advocacy* to mean promotion by professionals with expertise

81

knowledge, experience and positions to help ordinary local residents such as herders. Unlike advocates, activists do not necessarily speak for or in support of ordinary people but rather for their own interests. While advocacy was a primary characteristic of the ORM to begin with, friction between donors and activists inspired some of the movement leaders to take a community-based approach which resulted increasingly in the adoption of features of activism later on. In other words, the process was more or less the opposite of how Byambajav puts it. Advocacy led to activism, not the reverse. However, the ethnography also warns us that it would be misleading to make a stark division between advocacy and activism, as both function as different tactics available for protestors to deploy, often simultaneously.

This chapter also shows how the advocacy group started to produce and distribute discursive resources that articulated problems relevant to mining, environment, wealth and policy, and how those discursive resources established a right and recognisable way to protest. For example, with regard to protest movements, Orgilmaa Zundui-Yondon, a member of the family alliance of three gold mining companies, Cold Gold Mongolia, Odod Gold and Uuls Zaamar, said that after the windfall tax and the 'Law to prohibit mineral exploration and mining operations at headwaters of rivers, protected zones of the water reservoir and forest area' (known as the 'law with the long name') in 2009, many Mongolian people were transformed into potential 'movement persons' (*khödölgöö-nii khün*). As I find from my interview with her, people did not have to have a membership and movements did not have to be registered to form a movement (*khödölgöön*). There was no division between someone who was a member of a movement and someone who was not. Every request of any individual local person – such as student tuitions and fixing motorcycles – was backed up or secured with possible individual or group demonstrations against mining companies. Here, first, these discursive resources and the right to resist gave some leverage to people against mining companies and a chance for some small share of the wealth. Second, it does not seem as if there is a single giant 'movement', but rather a recognisable way to protest at the heart of the mobilisations. I take these points as an entry to this chapter to account for the series of mobilisations against gold mining companies and state (government and parliament) policies.

Regarding the first point about discursive resources and the right to protest, the first section of this chapter shows how from the start the popular mobilisation as an advocacy group produced and distributed different discursive resources; promoted the right to resist mining companies and neoliberal policies; and established recognisable ways to protest with which local people might be able to stop mining operations or extract some wealth from mining companies or the state. Regarding the second point about the singularity of the movement, I must note that my depiction of the success and power of those popular mobilisations does not intend to show as if there is a single movement. In this and also in the next chapter on protest movements, I examine the formation and dissolution of different advocacy groups and movements with different sizes, names, memberships, interests, purposes and methods of protest, and links along with the changes of emphasis and rhetoric in the process of constant formation and dissolution of the groups, all of which cannot be reduced to one thing and considered as one single movement. For this reason, none of the existing generic terms used to identify those popular mobilisations, such as populist, nationalist (*ündserkheg*), environmentalist (*baigali orchny*), civil society (*irgenii niigmiin*), local (*oron nutgiin*) and river (*gol mörnii*) properly captures the fluid, mobile and multifunctional features of those popular mobilisations. Each of these terms depicts a particular emphasis of certain mobilisations that last for a certain period of time, but fails to capture above-mentioned links and changes of emphasis and rhetoric and so the mobilisations as a whole.

The final point that I bring into consideration in this chapter is a discussion of power relations. As mentioned in the Introduction, in most of the works in the field of mining, resources, environment, activism, governance, state and capitalism, the protestors who fight against powerful international corporations are often powerless, frontier, minority, tribal, indigenous and local populations. On the other side, the corporations are often transnational and foreign (cf. Chapter 3). Chapters 4 and 5 show that this is not the case for these movements in Mongolia. Regarding this point, Anna Tsing (2005, 3) notes that 'it has become possible for scholars to accept the idea that powerless minorities have accommodated themselves to global forces'. She rejects this and turns the statement around to argue that 'global forces are themselves congeries of local/global interaction' and 'illustrates friction of global connections'. Similarly, mobilisations in Mongolia discussed in this chapter are powerful forces that shape mining companies, resource economy and politics. My ethnography of the series of mobilisations in Mongolia suggests that the local protestors cannot be reduced to a conventional presentation of the powerless; they can be powerful in different ways. This is partly because, in Mongolia, the local populations facing environmental problems are not ethnic minorities, tribal communities or indigenous groups of different races or even ethnicities.

Moreover, in many cases of environmental protest in Mongolia, it is sometimes the local government and authorities who start, advocate and fund strikes and movements against mining companies. Movement leaders often run for the parliamentary elections, and some become parliament members. They lobby politicians and laws to stop mining-caused environmental damage. Consequently, the state (government, parliament and sometimes the president) authorities often intervene into the power relationship between mobilisations and companies. For example,[1] Munkhbayar Tsetsegee, who founded the ORM, in addition to his experience with several parliamentary elections, is the winner of the internationally renowned Goldman Environmental Prize of 2007, and was named a National Geographic Emerging Explorer of 2008. Taking his achievements into consideration, the conventional presentation of the powerless and peripheral whose voices are not often heard in the national and global arenas is not necessarily the case in Mongolia.

Furthermore, in the literature I mentioned in the Introduction, there is a general depiction of the nation-state's stance to authorise the expansion of the extractive industry to improve the national economy. For this purpose, many nation-states tend to prioritise the extractive sector and national economy over the well-being of the local population and environmental destructions.[2] However, in the nation-state of Mongolia, including different rulers, politicians and policymakers in parliament and government, never continuously and sustainably promote and prioritise the extractive industry sector over the local population and environmental destruction. For example, the two main parties had a clear policy distinction regarding the extractive sector. Since the 1990s, the Mongolian Democratic Party (MDP) continuously showed enthusiasm and effort – especially when they ruled the country from 1996 to 2000 and 2012 to 2016 – to speed up democratisation, free marketisation, privatisation, foreign investment and the business of vast national reserves in the global market. Meanwhile, the Mongolian People's Party (MPP) has been conservative, careful and hesitant to make significant moves toward neoliberalisation (see also Rossabi 2005; Addleton 2013). As such, extractive industry projects are not continuously and uniformly prioritised in Mongolia. Consequently, some pivotal political decisions, laws and regulations to protect the environment, and the entitlement of the local population and local authority often bring detrimental impacts in the extractive sector (see Chapter 3).

Emergence of Advocacy, Discursive Resources and a Right to Protest

As we have seen in Chapter 1, in the 1990s and early 2000s, the immediate actions of the president Ochirbat Punsalmaa and the rule of the Democratic Party government served to urgently develop the mining sector and appealed to foreign investors to assist the national economy (see Chapter 1). As a result of the development of mining, some people started to experience environmental degradation and lack of natural water. One of the first incidents to become public was the case of Ulaan Lake (Ulaan nuur) and the Ongi River (Ongi gol).

In the late 1990s, many herders in the south of Mongolia started migrating due to the lack of water. Many of them were herders who lived along the Ongi River and Ulaan Lake in the eight *sums* (sub-provinces) (Bulgan, Mandal-Ovoo, Saikhan-Ovoo, Bayangol, Taragt, Züünbayan-Ulaan, Uyanga and Arvaikheer) of three *aimags* (provinces) (Ömnögovi, Dundgovi and Övörkhangai). Ulaan Lake started to shrink, and its primary source of water, the Ongi River, vanished by 2001. The gold mining left about 60,000 people – most of whom are herders – and a million head of livestock of eight *sums* along the river with no source of water (see also Snow 2010; Upton 2012, 240; Byambajav 2015, 93). The migration of herders created conflicts in the use of pastureland in the neighbouring *sums*. Bayarsaikhan Namsrai, a former school teacher and a newly elected chair of the Citizens' Representative Meeting of Bulgan *sum* of the Ömnögovi province, where Ongi River connects to Ulaan Lake, decided to meet officials of the neighbouring *sums* to discuss the problems of herders' migrations and pasture conflicts. When she met officials of the neighbouring *sums*, she discovered that many herders of those *sums* were also migrating due to the lack of water. There, she met Munkhbayar Tsetsegee in Saikhan-Ovoo, Dundgovi, who was at that time the chair of his *sum* Citizens' Representative Meeting, and other officials from the rest of the eight *sums*. Munkhbayar says that their local herders had been turned into 'ecological migrants' (*ekologiin dürvegchid*). Herders migrated some hundreds of kilometres to different provinces. However, Munkhbayar's focus was not only on those herders who migrated but also on those living in the *sumyn töv* (*sum* settlement) where the local government, school, hospital and all other services are. Many *sumyn tövs* are on the side of the Ongi River, and the natural water of the river is the main water supply for residents. One of those *sum* settlements is Saikhan-Ovoo, where Munkhbayar and his family lived. *National Geographic* (2008) portrayed a bleak situation: 'Desperate for drinking water, Munkhbayar's family and neighbours dug wells. However, groundwater was so contaminated that dozens of local children suffered serious liver damage. Munkhbayar's son was taken ill, and his mother lost her life'.[3] As such in the meeting of *sum* officials the lack of water, health care, the migration of herders and pasture conflict became the initial basis of the discursive resources which further generated rights to, and methods of, protest.

In the meeting of the *sum* officials, Munkhbayar suggested driving along the river course to its source to discover why it was drying up. From 23 to 30 September 2001, a group of local leaders including *sum* governors, chairs of Citizens' Representative Meetings, three journalists and a lama from the Saikhan-Ovoo monastery organised a trip along the river course and travelled to the place where the river commences in Uyanga (Amarsanaa and Bayarsaikhan 2007, 17). During the trip, they discovered the presence of more than thirty open-pit gold mines, and most importantly the unlawful and destructive nature of their operations (Amarsanaa and Bayarsaikhan 2007, 24). In Uyanga, close to where the river starts, they discovered a mining operation of Erel (described in Chapter 3). Erel made three dams to close the river in order to use its water for gold extraction. Those three dams closed the flow of the river water and left local people and livestock along the river with no source of water (see also High 2017, 56). They met the mining company and asked them to let the river water flow. The company responded that they could only open one of the three dams. The company's refusal led the group of locals and others to start a movement against Erel and other mines along the river.

Upon their return, on 18 and 19 October, they organised a meeting in Saikhan-Ovoo with local people and officials of the eight *sums* in order to inform them of the reason why the river dried up. In the meeting, they decided to establish a movement to protect the river. They called the movement *Ongi golynkhon* (People of the Ongi River), also known as the Ongi River Movement (ORM) (see also Sneath 2010, 262; Byambajav 2015, 93; High 2017, 55). The movement had 34 founding members: three *sum* governors; four chairs and one secretary of the *sum* Citizens' Representative Meetings; three government officials from the Ministry of Environment, including an international consultant from Japan; four local environmental officers; three media persons; two persons from the Gobi Regional Economic Growth Initiative (Gobi Initiative) Programme, managed by Mercy Corps and Pact and funded by the United States Agency for International Development; two monks from local monasteries; the mine director of Erel company; and other people such as the local school director, cultural centre director, weather forecast department director, tourist camp director, and archival worker and elders' association chair (Amarsanaa and Bayarsaikhan 2007, 17–19). While many of these elites might claim to be herders or to have herder ancestries, there was no one directly involved in pastoral subsistence.[4] As such, the movement was initially based on the unification of local government institutions – except the monastery and the association of elders. In addition to those local elites and leaders living in the region, some of the founding members were local people who moved to and were based in Ulaanbaatar, Mongolia's capital. They were the main allies for the movement to link this rural place to the nation's capital. The movement also had a board of 11 people, three of them governors and five of them chairs of the Citizens' Representative Meetings from the eight *sums*, plus another mine director and a professor from the Mongolian University of Agriculture. Munkhbayar was appointed leader of the movement. Regarding funding, each of the eight *sum* governments decided to donate 100,000 MNT in initial seed money. There were also other donations from the founding members, individuals and organisations. The network, influence and expertise of the founding members and board members helped the movement to immediately gain publicity through different media sources, such as a 13-minute television documentary on the Mongolian national television channel (Badamsambuu 2007, 126), a 50-minute radio programme that was broadcast on national radio, and reports in local and national newspapers (Amarsanaa and Bayarsaikhan 2007, 23, 28).

By talking to Munkhbayar I discovered that many of those members from outside the region were initially from these localities but had moved to and settled in Ulaanbaatar. To bridge the rural and urban, Munkhbayar suggested that those who lived in Ulaanbaatar should establish a council for people from Saikhan-Ovoo, known as *nutgiin zövlöl* (local homeland council or homeland association) (see also Sneath 2010, 258). *Nutgiin zövlöl* is a non-governmental organisation of people from the same homeland, who reside in the capital of the country or secondary provincial towns and other urban regions. The purpose of the council is to bring together politicians, businesspeople, artists, athletes and others with important positions to act for the well-being and development of the home locality, by informing, advertising, promoting and fundraising. Therefore, it was important for the movement to have people from the local homeland council in Ulaanbaatar. For example, one of the people to bridge the rural–urban gap for the ORM was Dr Chandmani Dambabazar, a water engineer from the University of Agriculture. Chandmani was born and raised in Saikhan-Ovoo, was a founder of the Saikhan-Ovoo local council in Ulaanbaatar, and became a board member of the ORM. Other founding members representing the central government from the Ministry of Environment, gold mining companies and media also had important roles: to enlarge the movement and attain recognition in the short term.

The ORM was also successful in forming collaborations with international donor organisations. Only a few months after the establishment of the movement, a German political foundation called Konrad-Adenauer-Stiftung (KAS) contacted the ORM and offered support to run the movement and establish a grassroots civil society (see also Byambajav 2015, 93). KAS is named after the first Chancellor of the Federal Republic of Germany and is closely associated with the Christian Democratic Union of Germany. The foundation supports education on freedom and liberty, peace and justice. Dr Peter Gluchowski was the regional director of KAS in Mongolia. He heard of the movement from his Mongolian employer B. Battuvshin, who witnessed the process while he was working as a teacher for a workshop on the local governance in Saikhan-Ovoo in autumn 2001 (Battuvshin 2007, 115). The purpose of KAS – to support civil society, democracy and justice – found in the ORM a perfect case to promote. For this purpose, members of KAS, including the regional director Gluchowski and his wife, travelled to the south of Mongolia to participate in the second meeting of the movement, which was on 6 April 2002 (Buyantogtokh 2007, 89–90). The contribution of KAS to the ORM was not only financial and technical but also conceptual. Michael Henke, invited by Gluchowski and funded by KAS to assist the ORM, made enormous contributions through his design of annual action plans of the movement and the education of movement members on subjects such as democracy, civil society and the state (see also Amarsanaa and Bayarsaikhan 2007, 27–33, 46–52). Henke was an experienced activist and mentor. Each year from 2002 to 2006, with the support of KAS, Henke visited Mongolia to teach and help to design the action plan and manage the movement. Amarsanaa and Bayarsaikhan wrote that Henke's teaching and guidance were 'eye-opening' (Amarsanaa and Bayarsaikhan 2007, 28). In other words, from the start the movement was not straightforwardly a grassroots one. Bayarsaikhan later told me that in Henke's classes she learned many Western concepts such as the civil (legal), liberal and welfare states, civil society, and the role of people in building a modern civil society. In the edited volume, she deliberately included some illustrations and the contents of what Henke taught the movement members (Amarsanaa and Bayarsaikhan 2007, 30–3). Movement members respectfully call him *bagsh*, which means teacher or master in Mongolian. This respect (see also Dulam 2006) places Henke and KAS in a position to influence, lead and mentor the movement. The most important thing the movement members learned was the systematic progression of actions, starting from peaceful resistance such as letter writing campaigns, collecting signatures and organising rituals and events, and eventually moving on to extreme and hard actions such as court appeals. In fact, the ORM tried almost all of those actions gleaned from Henke's workshops (see also Chapter 5). It must be noted that Henke's mentorship helped to recreate the understanding of the state as something that is high, distinct and respected (see also Dulam 2006), and split the deified and secular forms of the state (see Chapter 5). According to Bayarsaikhan, in Henke's classes, she first realised that she could shape the state by critiquing, complaining, demanding or even fighting against the political rulers. As a result, the movement expanded its resistance and fought against the state (government, parliament and the president; see Chapter 5).

In the following years, in addition to KAS there were many other donor organisations and companies that promoted the ORM, such as Merci Corps, World Bank, Soros Foundation, The Asia Foundation and Ivanhoe Mines (Amarsanaa and Bayarsaikhan 2007, 50, 54). According to Bayarsaikhan, funds for travel, field trips, visits, workshops and more were generously and frequently offered from different donors and companies. The funding sources required movement members to submit project proposals one after another (see also Addleton 2013, 39–40, 64–5). Among them was TAF, which offered funding from 2004 to promote river movement activities, and became the funder of the ORM after KAS. Layton Croft, who was director of TAF in Mongolia and who later became the director of the Canadian Ivanhoe Mines Mongolia, claims that 'ORM was not a government puppet operation or an international donor-initiated organisation; rather, it was the first and genuine local movement'. Moreover, he states, 'to help the movement to continue, TAF had to fund even its daily operation costs which foreign non-governmental organisations tend not to do' (Amarsanaa and Bayarsaikhan 2007, 118). With this funding, in 2004 the ORM opened an office in the centre of Ulaanbaatar and employed five people to work there (Amarsanaa and Bayarsaikhan 2007, 53). Yet, Croft's depiction of the movement as 'first' and 'genuine' may well be the way he needed to view and present the movement to justify his assistance. Unlike Croft's depiction, the last section of this chapter tells the opposite, which shows friction between local and donor interests.

What is clear is that from the start the ORM was not simply made up of herders and grassroots, peripheral, tribal or indigenous groups that represent a local community in the model of the dual and triad relationship of local community, state and mining company, as I noted elsewhere. Rather, the movement was not limited to the remote regional territory and network. To establish the movement, local leaders, such as Munkhbayar, Bayarsaikhan and others, intended to bring together people with different backgrounds, experiences, expertise, positions, residences and interests, which is evident from the list of the founding members and board members. These 'actors play an important role', as Anthony Bebbington et al. (2008, 2892) describe, 'in keeping movements "moving" – by maintaining debates, supporting events nurturing leaders and sustaining networks during those periods when movement activity has slowed down'. Therefore, the establishment of the ORM was the formation of a strong local advocacy group with good regional, national and international connections and with the capacity to produce different discursive resources such as those on the state, civil society, and rights and ways to protest. The range of membership helped the movement to create discursive resources that became deployable in diverse settings. More precisely, those discursive resources were a product of an intention in the working group to peacefully and jointly find a solution to local environmental problems through means other than fierce resistance, which nevertheless occurred in the later years of the movement.

Nationalisation and the Internationalisation of Activism

By 2003, Bayarsaikhan recalls that her group had started to view the drying-up of the Ongi River as a 'failure of the state' (*töriin buruu*), despite almost all of the founding members and leaders of the ORM having high positions in the local governments. By *tör* or the state she was referring to the ruling institution, namely the government, parliament and the president and their political decisions and legislations. Moreover, Bayarsaikhan says their obligations in the local governments to implement central government policies to promote mining were in conflict with their intention to protest against those policies. Therefore in 2003 Munkhbayar and Bayarsaikhan left their jobs and were followed by several others in the next few years. Their abandonment of their positions in local government made many of them into full-time activists in the protest movement rather than advocates in the working group. Here, we see a significant change of emphasis and rhetoric. The emphasis on the mining-induced environmental damage shifted to an emphasis on policy. The focus of their rhetoric changed from a critique of mining companies to a critique of the state.

In the midst of the 2004 parliamentary election campaign, the ORM leaders organised a protest march alongside the Ongi River to persuade local people to vote for candidates who would work for the protection of the river. Twenty-four people spent about four weeks walking from the mouth to the head of the river, holding public meetings with the local herders along the way. As a result, some candidates joined the march and proclaimed their intention to advocate for the movement (see also Byambajav 2015, 94). For instance, R. Raash, D. Lundeejantsan and Z. Enkhbold joined the march. Bayarsaikhan suggested that the action plan of those candidates to support the movement, environment and well-being of the local people became an important criterion to win the election (see also Amarsanaa and Bayarsaikhan 2007, 76–7). After winning the election, R. Raash established a parliamentary lobby group to support the movement as planned, resulting in a continuity of advocacy at the level of parliament. The lobby group in parliament helped the movement leaders to transform the issue from a local struggle to something of importance to the nation in what might be called a process of nationalisation.

Although the coalition of the Erel company's Motherland Party (*Ekh oron nam*) and the Democratic Party saw a tremendous victory and established a government (see Chapter 3), this has nevertheless produced favourable conditions for advocates and activists to mobilise within. For the political opponents who fought against the Motherland Party, Erel's environmental catastrophe and the claims of the movement against Erel served to weaken the power of the Motherland Party. The lobby group in parliament became a significant force to fight against the power of the Motherland Party. The discussion of the mining catastrophe in the national-level political debates – and the importance of the movement within the debate – helped the local movements to nationalise their problem. It was a marked achievement for the movement to generate support for the local people in the fight against mining companies. Amarsanaa and Bayarsaikhan write that this accomplishment made local people realise that they could influence the rule of the state (*töriin erkhiig barikh*) (see Introduction).

With the support of the local people, local governments, their allies in Ulaanbaatar, and some politicians in parliament, the ORM dramatically expanded in the next few years. The growth of the movement progressed by transforming seemingly local problems into a national concern. This came about in two ways: First, politicians used the environmental problems to make an argument against the ruling Motherland Party in the coalition with the democratic union. Second, the movement was able to present the local problem as a national problem by involving people representing almost all regions of the country. Within two years, the movement established branches in the eight *sums* along the Ongi River, which included 1,400 members and 11 member organisations. The number of member organisations increased further to 26 by 2005. Member organisations of the movement were local governments, hospitals, kindergartens, secondary schools, research institutes, monasteries and private companies (Amarsanaa and Bayarsaikhan 2007, 48, 56). Among the most critical impacts of the movement was their contribution to the establishment of other similar river movements across the country. According to Munkhbayar, as early as 2003 he realised that the problem was not confined to the Ongi River; it also affected many other rivers in very similar situations. Eventually, the ORM received requests for help by people from around a dozen regions. Members of the ORM were invited to different provinces to distribute and establish a right and recognisable way to protest, by sharing their experiences and organising, mentoring and managing local movements across the country. In 2005, 11 local river movements representing 14 different provinces[5] (out of 18 in total) united and formed the *Mongol nutag minu evsel* (My Mongol Homeland coalition). The coalition was also known in English as 'Homeland and Water Protection Coalition' (HWPC) or 'Mongolian Nature Protection Coalition' (MNPC) (Amarsanaa and Bayarsaikhan 2007, 53, 58, 83–4; see also Byambajav 2015, 94–5).

The actions of the movements against mining operations were publicised as peaceful and non-violent (*taivan*). The movement members visited mines and mining-damaged sites and sent letters of resistance to mining companies. Many members also visited local herder households and marched along the river course to collect signatures of the local people and to organise meetings and workshops. Members of the movement collaborated with national and international environmental scientists and scientific institutions to research environmental damage, participated in conferences and meetings, and produced television and radio programmes, newspaper articles and interviews. To help the environment, movement members planted berry trees along the Ongi River course and created artificial rains by cloud seeding.[6] The members of the movement introduced a course in the secondary school curriculum and published textbooks on ecology and the environment to inform and educate the younger generation. To attract attention and influence national politics, they sent letters to the president, prime minister and parliament members, and organised meetings with election candidates, politicians and political parties. By the end of 2003, however, actions of the movement shifted the emphasis of their force to pressure mining companies. They requested mining companies report to the local authorities and residents, lobbied local governments to make decisions to stop the operation of mines that failed to rehabilitate the land, and began to threaten the corporate reputations of some companies. As a result, some mining companies rehabilitated mined sites, while some others had to cease their mining operations (see Chapter 3) (see also Amarsanaa and Bayarsaikhan 2007, 48–53, 58, 82, 86). The resultant return of the river in the territory of seven *sums* (out of eight) and 430 kilometres out of the total of the 435 kilometres' length became a symbol of the success of the movement. From 2005, 'on behalf of the state' (*töriin neriin ömnöös*) and with the help of the Center for Human Rights and Development, the movement initiated criminal proceedings and lodged a complaint against Erel to compensate local people for environmental damages (see also Amarsanaa and Bayarsaikhan 2007, 58, 60, 62). As a result, in the following year, the movement successfully caused the closure of 35 of 37 gold mines along the course of the Ongi River.

Here, what I call nationalisation processes highlight how the central government and politicians in parliament participated, collaborated and supported the movement, which suggests the opposite of what is described in most of the pre-existing literature on the subject, as I mentioned in the introduction of this chapter. Politicians in the central government and parliament did not uniformly support mining and the national economy; instead, some supported the river movements. The political climate created advantageous conditions for the movement to expand. In other words, it was not only the power of the movement but the condition of the national politics that allowed the movement to be successful.

In addition to the nationalisation processes described above, processes to involve different people and organisations internationally and to become known to the international audience which can be termed 'internationalisation' occurred in two steps: first, by way of collaboration with donor organisations to establish and develop the movement (see above); and second, by winning internationally prestigious prizes and titles. Munkhbayar received the 2007 Goldman Environmental Prize for uniting 11 river movements from different parts of the country, closing 35 mines, attracting the attention of parliament and the central government and successfully nationalising the movement[7] (see also Snow 2010). In 2008, Munkhbayar was also named as the National Geographic Emerging Explorer. Collaboration with donor organisations were the main factor for the movement to become internationally known. Layton Croft, director of TAF in Mongolia (2003–5), nominated Munkhbayar for the prize. Croft began his career as a Peace Corps volunteer in Mongolia (1994–7). He later joined the Mongolian office of Pact, a large US government and intelligence organisation, where he was the Program Director for Information Systems for their Gobi Regional Economic Growth Initiative/Mongolia (1999–2002), working for Pact-Mongolia in an alliance with Mercy Corps and the United States Agency for International Development (Snow 2010). Given such experience, Croft was well informed about the river and the river movements from 1999 (Amarsanaa and Bayarsaikhan 2007, 117). He was in an excellent position to negotiate between transnational mining corporations and environmental movements.

Although the movement leaders risked their authority in the movement, as I will illustrate in the following section, the internationalisation of the movement – with the support of TAF and other donors – helped the movement leaders to gain recognition and power in national and international scales (for internationalisation of movements in South America see also Bebbington et al. 2008, 2901). In other words, TAF created a powerful position for the movement on an international scale. Chapter 5 will show how Munkhbayar further used this position to fight not only against mining companies but also against the government. However, while these nationalisation and internationalisation processes empowered the movement, it also made the movement fragile. Bebbington et al. (2008, 2892) note that such an internationalisation is a source of both weakness and power and that for movement leaders it is an immensely difficult feat 'to hold the movement process together around a shared agenda and vision'. The same also applies to the unity of river movements and their collaboration with donor organisations in Mongolia. The national and international expansion of river movements put the success of the movement in jeopardy.

The Friction between Donors and Movement

Here, to examine the relationship of the movement and donors, I am using Tsing's (2005, 4) argument about friction in the confluence of the universal and global on the one hand and the particular and local on the other, which together creates what one might describe as capitalism and neoliberalism. The friction caused another change in emphasis: the movement's leaders started to emphasise the risks and problems of donor funding and advocacy.

In 2005, TAF director Croft – who was the primary funder and supporter of the movement – became the executive vice president for corporative affairs for Ivanhoe Mines, operator of the vast Oyu Tolgoi gold and copper mine in Ömnögovi, in the province where the Ongi River reaches Ulaan Lake. Croft's new job was a surprise for Munkhbayar and his colleagues in the movement. They had to consider how such a shift could happen and what it meant for the movement. This new position made the movement members realise that Croft had his own agenda at TAF. Both Munkhbayar and Bayarsaikhan said they realised that, while at TAF, Croft initiated and offered funding to implement different projects, but not necessarily to support the movement to fight against mining damage. At the same time, the increasing influence of TAF made Munkhbayar, Bayarsaikhan and other members question the leadership of the movement. Others also noticed this. For example, Jane Smith, a conservationist in a small NGO in Mongolia, says that TAF cut off the heads of the rural organisations by bringing them to Ulaanbaatar and giving them nice offices (Snow 2010). Although Munkhbayar noticed these problems, he decided that to receive the Goldman Prize, maintain international recognition and bring the voice of the local movement to the international level, provided enormous social capital for the future of the movement. Immediately after he received the Goldman Prize, Munkhbayar decided not to work with TAF; he explained to me that TAF was supporting mining and using the status of the movement and Munkhbayar as cover for the operation of the Ivanhoe Mines in Mongolia. Croft's intention to use Munkhbayar and the movement already appeared in some of his speeches and interviews. In an interview in 2007 Croft stated that the 'movement is not against mining, but against illegal mining' (Amarsanaa and Bayarsaikhan 2007, 119). Croft also said something similar in Munkhbayar's Goldman Prize profile video. In the video, he said, 'the key to Munkhbayar's success as a leader for responsible mining in Mongolia is that he has had the courage to acknowledge that mining could be a good thing for Mongolia, as long as it is done in a very open and participatory way'.[8] As Mette High (2017, 57) accurately points out: 'As a figure of international renown, Munkhbayar found himself becoming a valuable commodity for not only politicians but also the mining companies he was fighting against. The very people he opposed co-opted his agenda, and he became their key icon of Mongolia's grassroots consent to mining'.

The foreign concept of 'responsible mining' (*khariutslagatai uul uurkhai*), successfully introduced in Mongolia by TAF and other donors, was one of the main ways to commodify and co-opt Munkhbayar and the river movements. For Munkhbayar, the concept was an alternative discursive resource for the movement to negotiate and collaborate with mines rather than to resist them. Some found the idea of 'responsible mining' to be a better way to negotiate with mining companies and to receive funding, while some others, including Munkhbayar, proposed fiercer methods to fight against mining. He suggested the use of force, violence and arms for self-defence if necessary. He also suggested that the movement should not receive any more donor funding, and claimed they should lobby parliament or the government to pass an official declaration to prohibit mining in water sources and forest areas, an idea that later resulted in the bill 'Law to prohibit mineral exploration and mining operations at headwaters of rivers, protected zones of the water reservoir and forest area'. Leaders of the 11 river movements in the coalition came to an agreement on the above decisions. However, in April 2008, disagreement dissolved the coalition. Five of the movements remained in the MNPC and continued their collaboration with TAF. As Munkhbayar explains, those five movements felt strongly that they would require donor funding to continue, that they should never use force, and believed that it would be impossible to make parliament or the government pass a decision to prohibit mining in natural water sources and forest areas. They decided to continue their actions by creating a tripartite contract between the mining companies, local government and the local movements. This position of the five movements further incorporated the concept of 'responsible mining'. The remaining six movements continued their protest together and, on 4 June 2009, renamed themselves as United Movement of Mongolian Rivers and Lakes (UMMRL) (see also Simonov 2013).

Writer and activist Keith Harmon Snow visited and travelled in Mongolia in 2008 and investigated the river movements and their tricky relationships with TAF. In 2010, he published an online article where he states that the movement and donor relationships and the award of the prize was a 'Western deception'. He claims that TAF's agenda was not pure and not for the interests of the environment and people. Instead, he claims that TAF's agenda was to maintain control over the movements and use their success story to leverage €2.7 million (US$3,630,000) from the Dutch government. He also claims TAF had further intentions to shape Mongolian civil society in the interests of Western penetration and control and to forbid them from publicly protesting against mining companies or government policies in order to suggest a welcoming mining climate where companies comply with the environmental stewardship. Referring to Croft, Snow admits that his agenda was in service of a personal career move to become the Executive Vice-President for Corporate Affairs and Community Relations at South Gobi Resources, an Ivanhoe Mines/Rio Tinto megaproject, and an adviser for investor relations in Asia and corporate social responsibility for Ivanhoe Mines (Snow 2010). Moreover, Snow notes that 'Mongolian NGOs were expected to defer to TAF when dealing with the media, and they were compelled to sign contracts forbidding them from publicly protesting against mining companies or government policies. TAF also worked to determine and control the members of the boards overseeing the environmental coalitions that received TAF funds. Ultimately, river coalition members found they had no control over their own groups: TAF tried to maintain all control' (Snow 2010).

When I talked to Munkhbayar about what Snow had written about his relationship with TAF and Croft, he said Snow had been naïve to trust donor organisations. Bayarsaikhan said the same thing. They were sincerely grateful for TAF's support in understanding the situation of mining damage and the river movements' intentions to protect the environment. Yet both of them stated that they later realised why Croft was so enthusiastic to help the movement at TAF. The river movements' experience with TAF shaped the future of the movement significantly. At a minimum, the experience divided the united force of the 11 river movements and introduced an alternative way to negotiate with mining companies through the concept of responsible mining, while those led by Munkhbayar learned to not trust any donors. The mistrust encouraged the movement to self-fund, in order to protect themselves from the interests of donors and attempts to commodify the movement. The conflict between the river movements and TAF is what Tsing (2005) calls the 'friction' of local–global collaboration, as I mentioned before. She argues that their collaboration does not mean that the local and global have common interests but rather the collaboration maintains difference and friction at its heart (Tsing 2005, 246, 248–9, 264). Therefore, the underlying 'friction' in the collaboration creates new interests and identities, and new emphases, which are created by local environmental protestors while escaping from the global civil society (Bumochir 2018a).

Protecting the Movement from Donors

At the end of November 2015, when my colleague Byambabaatar and I first arranged to meet Munkhbayar in Ulaanbaatar, the first question he asked us was about the funding for our research. His main concern was that international donor organisations and funders could have different hidden agendas and at a minimum contribute to creating a repository of information which could potentially be used later against the interest of local people and the nation. There is increasing suspicion towards donors' agendas and advocacy, which might not mesh well with the protestors' purposes. These suspicions about donor agencies can also taint the reputations of the protestors. What can also be called into question are their motives and reputations: are they being paid to accomplish certain tasks or are they genuinely committed to protecting the environment? In Mongolia, many people suspect and reveal their concern about money-making and money laundering, the interests of politicians, donors and transnational corporations. With funding and donor advocacy come great risks of the misrepresentation of local protest movements; they can be made (or appear to be) puppets for larger international and national political institutions and agents, and the protestors' original interests and intentions can be multiplied, split and obscured (see also Bumochir 2018a).

Therefore, Munkhbayar decided that the safest way to manage a protest movement is to fund its activities with politically unproblematic and secure financial sources, which became a new emphasis of the movement. Their rejection of donor funding left those movements and local herding families who supported them in a very difficult environment to succeed within. To accomplish such a vital and immediate requirement and to succeed in a very difficult environment, Munhbayar suggested that they plant berries (sea buckthorn) along the course of the Ongi River, something he learned from his mother. He took us to a beautiful small garden with bushes and berry trees alongside the Ongi River near the Saikhan-Ovoo *sum* settlement his mother built in the late 1980s and early 1990s. The creation of such gardens in the area has multifunctional advantages both for the movement and for local people to succeed. First, they intended that this would maintain the water level of the river. Second, according to the legislation permits, any land with cultivated trees and bushes can be privatised by those who cultivate it; local herders can own land along the river, which prevents the loss of local pasture around the river to mining companies. Third, they can profit from the berries by selling them, and the movement can be self-funded in the future by using the income from berries.

Planting berry trees is not the only solution initiated by Munkhbayar to self-fund the movement. He also founded a herd for the movement, which was a relatively new initiative in comparison to the planting of the trees. At the board meeting of the ORM in November 2015, he presented his idea to create a herd to fund the movement. The easiest and quickest way to make it happen was by collecting donations from local herders. In the meeting, members representing five different *sums* along the river were present. There was general agreement at the meeting that the idea was excellent. It would make the movement financially sustainable and protect it from any ambiguous intentions of donors. Munkhbayar recalls that he was surprised to see the members react so quickly: within a week, every *sum* had decided to contribute around 80 sheep. When I visited him in December 2015, he had over 200 sheep for the movement outside his yurt. Two young men, both herders and members of the movement, decided to help by voluntarily taking care of the river movement herd.

Munkhbayar's initiative to self-fund the movement is completely different from how Bayarsaikhan describes the earlier stages of the movement and their informal policy on the acceptance of funding. As he admits, when Munkhbayar started his protest in the early 2000s, he did not know much about concepts such as civil society or grassroots politics, nor the consequences of donor advocacies. Therefore, he welcomed extensive support from different international donor organisations; namely, KAS and TAF. It could be argued that we are now experiencing a new phase of the environmental movement, with attempts to divorce itself from foreign donors and Western advocacy. The consequences of foreign interference sparked the separation and caused the ORM to take on a more grassroots approach (see also Upton 2012, 244).

Byambajav's (2015, 97) conclusion about the increasing transformation of local movements into non-community based foreign advocacy organisations may be true for those in some river movements who accepted the donor-initiated concept of 'responsible mining'. However, it is no longer valid and relevant to explain those who sought to escape from donor advocacy and attempt to refocus upon grassroots activism. As I argue elsewhere (Bumochir 2018a), these environmental protestors are reluctant to be associated with even the name 'civil society' (*irgenii niigem*) because, for them, 'civil society' organisations in Mongolia often import Western concepts and practices, which are unacceptable to the protestors. The river movements attempt to appeal not to the West, but to the past nomadic history and 'traditional' concepts regarding the environment and state protection, which have salience in many political contexts in Mongolia. Such concepts will be the focus of the next chapter.

Notes

1   A well-known singer and a founding member of Fire Nation coalition who became an MP in the 2016 parliamentary election. He named the coalition as *Gal Ündesten*, which means 'Fire Nation'. The coalition consisted of different environmental movements across the country.

2   This general depiction is not accurate in some cases: for example, in the work of Diego Andreucci and Isabella M. Radhuber (2015) in Bolivia, where many of the advanced laws and regulations to protect the environment and local population passed.

3   For more information see www.nationalgeographic.org/find-explorers/tsetsegee-munkhbayar

4   However, many of these people have livestock or experience in herding, which is typical all over rural Mongolia.

5   Those include Khangiltsag River movement from Uvs; Salkhin Sandag movement from Govi-Altai; Call of the Khongor Homeland (*Khongor nutgiin duudlaga*) movement from Bayankhongor; Sacred Suvraga (*Ariun Suvarga*) movement from Arkhangai; Toson Zaamar-Tuul River movement from Töv; Khüder River movement from Selenge; Onon-Ulz River movement from Khentii; Native (*Uuguul*) movement from Ömnögovi; Masters of the Khövsgöl Lake (Khövsgöl dalain ezed) movement from Khövsgöl; People's Envoy (*Ardyn elch*) movement from Selenge; and the Ongi River Movement.

6   Cloud seeding is the process of combining different kinds of chemical agent – including silver iodide, dry ice and even common table salt – with existing clouds in an effort to thicken the clouds and increase the chance of rain or snowfall. The chemicals are either shot into the clouds or released by flying near and into the clouds (Kirkpatrick 2018).

7   For more information, see www.goldmanprize.org/recipient/tsetsegee-munkhbayar/

8   For the full video, see www.goldmanprize.org/recipient/tsetsegee-munkhbayar/

5
The De-deification of the State

According to Fernando Coronil (1997, 8), in Venezuela the state was construed as the legitimate agent of an 'imagined community' (Anderson 1983). Moreover, Coronil shows how the state was deified and 'state representatives, the visible embodiments of the invisible powers of oil money, appear on the state's stage as powerful magicians who pull social reality, from his public institutions to cosmologies, out of a hat' (Coronil 1997, 2). As I have demonstrated elsewhere (Bumochir 2004, Dulam 2009), there is a history of the deification of the state and its power in Mongolia (*töriin süld*), engaging Chinggis Khan and the supreme deity *Mönkh Tengger* (Eternal Heaven). However, the construction of the state generated by the activists discussed in the previous chapter was not a deification but the opposite: what I call de-deification. More precisely, ethnography in this chapter shows how activists created a division between the deified ideal of the state and the actual form such as its institutions, structure, system and authority (for the two forms see also Abrams 1988; Taussig 1992; Navaro-Yashin 2002). Here, I must note that de-deification does not work precisely *against* deification. Instead, the separation made between state institutions and deified forms of the state helps to free state rulers, institutions and law from the culturally salient legacy of the deified power of the state (*töriin süld*), and from customs and taboos that restrain action in respect of the state. Therefore, once the state system, institutions and agents become free from deified state power, then decisions, policies, laws and regulations are rendered open to criticism, attack and other forms of resistance against the state. Moreover, the separation of the two forms of the state achieved by this de-deification permitted activists to participate, involve themselves in and influence political decisions and legislation, and even to consider taking control of the state.

It would be misleading to assume that the de-deification of the state was simply the result of the transition to democracy (Lhamsuren 2006, 93). Caroline Humphrey and A. Hürelbaatar (2006, 265–6) located the phrase *törü-yi abuba* (take over the state or government) in a seventeenth-century text. They explain that 'in this period *törü* was something that could be created or established, handed over, corrected, administered, negotiated and discussed'. The ethnography I present below describes how activists construct such histories and actively engage in the creation, correction, negotiation and discussion of the state. Activists shape the state as something that is not far off, detached and superior from people but something that is available to them and can be taken over (*tör avakh*), ruled (*tör barikh*) and crafted.

Jürgen Habermas (1987) provides an account of the social and cultural resources drawn upon by social movement as they emerge. Anthony Bebbington et al. (2008, 2890) summarise Habermas's point and write 'social movements are apt to emerge when people's lifeworlds – their domains of everyday, meaningful practice – are "colonized" by forces which threaten these lifeworlds and people's ability to control them' (see also Habermas 1987; Crossley 2002; Edwards 2008). Habermas's depiction of the threatened 'lifeworld' is similar to activists' depiction of their local environments threatened by mining, and explanations for why they protest. Although Habermas was not interested in nationalism in this context, his description of the threatened 'lifeworld' and 'people's ability to control' helps to describe how and why activists mobilise nationalist discursive resources. Lowel Barrington (2006, 21) explains that 'in the minds of nationalists, the state, as a nation-state, exists for the benefit of the nation'. Therefore, if the nation's cultural identity is threatened, state policy must be adopted to protect the culture from the threatening 'other'. In a similar way, Munkhbayar explains that because the state failed to perform its duty of protection, activists had to step in to secure the 'lifeworld' they live in. Elaborating on Habermas and Barrington, I argue that the change in the emphasis and rhetoric away from local environment and livelihood protection to the protection of the territory and its people of the nation from external forces encourages nationalism within the movements. As I show in the last section of the chapter, this nationalism yearns for statism and aims to bring the state back or to 'make the state present' as Bebbington (2012b, 222) explains, or to bring some of the 'state provided services' (Bebbington 2012b, 222 cites Watts 2003 and Ferguson 2006) in order to protect its people, environment and territory.

The Law Behind the 'Long Name'

The use of arms to protect rivers was neither the sole nor main protest method for these movements. The primary and most effective method was to lobby the government, parliament and the president to pass a decision or environmental law to prohibit mining. From around 2005, the MNPC started to lobby the central government and parliament to pass a decision to prohibit mining at the headwaters of rivers, protected zones of the water reservoirs and forest areas. This was not possible for many years. As I described in Chapter 4, five of the movements in the MNPC did not find this plan to be realistic or accomplishable. After many years of failed attempts at lobbying state institutions, the activists came to blame the state for the issues. After attempting several different fight tactics, activists finally came to recognise failures of state regulations in the mining and environment. Both Munkhbayar and Bayarsaikhan narrated that in the initial stages of the ORM they did not blame the state. The sole purpose of the movement was to take the river back by fighting against the gold mining companies, particularly against Erel, the largest and the most destructive gold mining company. Almost all members of the movement blamed the mining companies, not the state. Bayarsaikhan says that it was only G. Badamsambuu, a journalist from the *National Broadcaster*, who argued that the state should be blamed (*töriin buruu*) for the environmental destruction. He accused the state – parliament, the central government and the president – for the following: First, the government issued mining licences in natural water and forest areas and created the risk of environmental degradation for many people. Second, the government failed to practise the rule of law to control the operation of mining companies to rehabilitate and safeguard the environment. Because the policy of the state institutions was initially central to the creation of the environmental damage, it should be state institutions that should be held responsible for finding solutions. This logic caused the river movements to eventually target state institutions, but an important question remained: How did they make the state listen to them? How did they make the state admit responsibility? To do this, Munkhbayar's and many of his colleagues' former positions and experiences in the local governments helped him to conceptually separate those who work for the state from the power of the deified form of the state to allow them to critique and blame.

Munkhbayar met different lawyers to accomplish this mission, including several famous ones; almost all of them told Munkhbayar that it is legally and ideologically impossible to make the state admit guilt. However, one of the lawyers he met was Dashdemberel Ganbold. Dashdemberel was the only lawyer who agreed that it was possible to place blame on the state. Dashdemberel helped him to pass the law to prohibit mining in the river and forest areas (the 'long name' law), and to amend the 'Environmental Protection Law' (*Baigali orchnyg khamgaalakh tukhai khuuli*) (Mongol Ulsyn Ikh Khural 1995). In July 2009, the Parliament of Mongolia passed the 'Law to prohibit mineral exploration and mining operations at headwaters of rivers, protected zones of the water reservoir and forest area' (*Gol, mörnii usats büreldekh, usny san bükhii gazryn khamgaalaltyn büs, oin san bükhii gazart ashigt maltmal khaikh, ashiglakhyg khoriglokh tukhai*) (Mongol Ulsyn Ikh Khural 2009). Munkhbayar complains that nicknaming the law as the 'law with the long name' was a political tactic to mask its intent. For him, the full intention of the law appears in its long name. Following his complaints, and to reclaim the law back from the 'long name' tag, I shall use the full and official name of the law in this chapter.

The idea of the 'Law to prohibit mineral exploration and mining operations at headwaters of rivers, protected zones of the water reservoir and forest area' started not from a bill but from a draft of an official decision to lobby, as I mentioned to start this section. According to Munkhbayar, the movement members eventually realised that it was extremely difficult to make any of the state institutions pass such a decision; to lobby such a law had a much better chance of success. In the year before the law passed, the movement intensely promoted the law both in the local-level and central-level governments. Munkhbayar and Dashdemberel met approximately 35–40 parliament members out of a total 76, along with the president, prime minister and the speaker. Also, in the course of one year, Munkhbayar and his colleagues personally visited about 1,000 households, most of which were made up of herders and their families along the course of the six rivers of the six movements. He briefed them on the environmental situation of those rivers and explained the importance of the law. After introducing and explaining the issue, Munkhbayar asked whether they would like to promote the movement. According to him, almost everyone responded in favour of support and often asked him how to promote the movement. Munkhbayar suggested three different methods for local herders to participate and promote the bill. They suggested sending letters, telegraphs or text messages; all of these were available immediately to herding families. Munkhbayar and his colleagues carried all of the facilities with them when they visited different families. For example, they helped them to write letters and telegrams and place them in envelopes. After each visit to a particular region, they brought all of the letters and telegrams to the local post office to send them. Text messages were the easiest, and Munkhbayar made sure that they sent text messages while he and his colleagues were present. In case people in the countryside did not have sufficient mobile phone credit, Munkhbayar and his colleagues brought different mobile phone operators' credits. They also had mobile phone numbers for all 76 parliament members, the speaker, prime minister and president. According to his calculation, in total, the families sent around 3,600 text messages to these politicians, 270 telegrams to the speaker, and 360 letters (with three copies each: one to the speaker, one to the prime minister and one to the president, for a total of 1,080 letters). Munkhbayar and his colleagues also produced a series of television documentaries called *Call of the Rivers* (*Goluudyn duudlaga*), which was broadcast on numerous television channels. All of these were funded by Munkhbayar's Goldman Prize of US$125,000, which was about 150 million MNT.

Parliament, the presidential office and central government could not ignore those letters, text messages and telegrams. Furthermore, the public had enough information about the damage done to different river systems across the country. Dashdemberel drafted the bill of the 'Law to prohibit mineral exploration and mining operations at headwaters of rivers, protected zones of the water reservoir and forest area'. The movement leaders chose Bat-Erdene Badmaanyambuu, a famous former national wrestling champion, to introduce the bill into parliament, which occurred in November 2008. Even though they had adequate support, they had to go through a series of negotiations with parliament in order to pass the bill. One item under discussion and debate concerned an article in the bill to enable citizens and NGOs to take legal action against the government for improper legislation. As a result of the series of negotiations with the members of the government and parliament, the movement was forced to eliminate the article to take legal action against the government. Nevertheless, later in 2010, Munkhbayar and his colleagues managed to include this article in the 'Environmental Protection Law' (*Baigali orchnyg khamgaalakh tukhai khuuli*) (Mongol Ulsyn Ikh Khural 1995, Article 32).

The bill intended to prohibit mining operations of river and forest areas. Article 4.1 of the law states 'Mineral exploration and mining operations are prohibited at headwaters of rivers, protected zones of water reservoirs and forested areas within the territory of Mongolia' (Mongol Ulsyn Ikh Khural 2009, Article 4.1). Then, Article 4.3 states

'The Government shall set the boundaries of the areas referred to in Article 4.1 of this law' (Mongol Ulsyn Ikh Khural 2009, Article 4.3). At the time the government set the boundary of the area at 200 metres. It included a list of mines and mineral deposits of strategic importance to the national economy, where most of the gold mines operate. In 2009, there were about 15 mines and mineral deposits listed as strategically important that were excluded from the law. These included Erdenet, a gold and copper mine that opened in 1974 as a joint Soviet and Mongolian government venture, and another much larger mine, Oyu Tolgoi (OT), a gold and copper mine venture of Rio Tinto and the Mongolian government, the construction of which was about to begin in 2010. Considering both the symbolic and financial importance of these mines for the nation and the national economy, Munkhbayar said that they had to agree that the law excluded those mines and deposits with strategic importance.

The law passed on 16 July 2009. One year later, in June 2010, the government revealed a list of 1,391 exploration licences, 391 operation licences and the operation of 242 gold mining companies (Il Tod 2012) that were now prohibited by the law, which directly and indirectly impacted the business of around 900 companies (Mongol News 2011b). The government of Mongolia had to pay compensation of at least MNT 647.3 billion (approximately US$460 million in September 2012) (Bold-Erdene 2013). As some officials complained, in addition to affecting hundreds of mining operations and licences and negatively affecting businesses of some hundreds of companies, the result of the law brought difficulties in the other sectors and the national economy. Of the 391 operation licences, 172 were owned by Mongolian companies, many of which invested in the other industries such as construction and food productions (see Chapter 3). Furthermore, many of those companies had bank loans, and stopping or failing to compensate them would have caused severe damage to many banks. Some estimates suggested that around 70,000 miners and engineers were at risk of unemployment (Mongol News 2011b). Therefore, immediately after its adoption, the law encountered strong resistance from many domestic and international mining companies, associations, investors and donors. The opposition against the law encouraged activists to continue their protests: they felt it necessary to press the government to enforce the implementation of the law and force mining companies to follow the law. In April 2010, the board of the United Movement of Mongolian Rivers and Lakes (UMMRL) decided to organise a series of actions to pressure the government and mining companies. UMMRL had two main ways to resist. One was to hold an armed protest; the second was to amend the 'Environmental Protection Law'.

Nationalist Movements against the State

In 2010, the leaders of the UMMRL decided to meet leaders of different protest movements. On 25 May 2010, UMMRL leaders recruited five other nationalist movements to continue and expand their resistance. By the end of the year, the new united movement of 11 different nationalist movements (*bülgem*), was officially reorganised as *Gal ündesten* (Fire Nation [FN]) coalition. Initially, there were five movements, including the following: *Mongol ulsyn ayulgüi baidlyn tölöö zütgeye* (Let's serve the security of Mongolia), established by some former military personnel; *Khokhirogchdyn negdsen kholboo* (United union of victims), formed by the victims and their family members of the mass violence that resulted from the parliamentary election on 1 July 2008; *Khökh Nuruut* (The Blue Ridge), an NGO established by famous singer Javkhlan Samand, who became a parliament member in 2016; and two other groups with the focus to protect the sovereignty of Mongolia, *Tusgaar togtnol evsel* (Independence coalition) founded and led by L. Tsog, former Minister of Law (1999) who became a parliament member in 2012; and *Tusgaar togtnolyn tölöo negdegsed* (Those united for independence) founded and led by O. Lkhagvadorj, former director at the National Centre of Construction and City Development (2008). These groups announced that they were united under the banner 'To save the country and the nation!' (*Uls, ündestnee avarya!*); they invited all Mongolians to join them. In the following months, additional movements joined. Those included *Chinggis khaany delkhiin akademi* (The world academy of Chinggis Khan), an NGO established by P. Davaanyam, who claims to be the descendant and successor of Chinggis Khan; and *Tenger ugsaa niigemleg* (Heavenly lineage society), established by N. Davaa, board member of the Anod Bank of Mongolia, to search for the burial of Chinggis Khan. Three additional movements later joined, bringing the total to 11. All of them shared the view that nation-state rulers violated and risked the *ündesnii yazguur erkh ashig* (the original interest of the nation), which demands that rulers have a concern for sovereignty and national security. In other words, leaders of these movements believe that a nation has an original or 'destined right' to have a politically sovereign state (Lhamsuren 2006). On 2 July 2010, leaders of the united movements sent a letter of requisition with eight points developed by each of the eight groups and demanded that the president, prime minister and speaker protect the interests of the nation.

The first point concerned the improper implementation of the 'Law to prohibit mineral exploration and mining operations at headwaters of rivers, protected zones of the water reservoir and forest area'. The second contested that the 10 per cent share of the Tavan Tolgoi coal mine that would be distributed to the people of Mongolia was insufficient. The third was addressed at the OT gold and copper mine contract between the Mongolian government and the Canadian Ivanhoe Mines company. The complaint argued that the contract violates the right of the nation to own its natural resources. The fourth complaint decried the increasing number of foreign employers in the mining sector, while the country continued to struggle with a high rate of unemployment. The fifth complaint was about corruption and accountability in government institutions. The sixth renounced the import of foreign products and use of US currency in the Mongolian market. The seventh complaint sought to protect the rights of victims of the gold mining pollution in Khongor. Finally, the eighth complaint was about the accusations of election rigging and the resultant violence on 1 July 2008. The letter demanded that the authorities respond, otherwise they planned to use other ways to resist, and the state would be responsible for any damages. None of the authorities responded to the letter. The movement waited for exactly 60 days after sending the letter, which was the statutory period for authorities to respond to complaints. On 2 September 2010, the movements' coalition, led by Munkhbayar, began their next action called *och khayakh ajillagaa* (operation to make a spark) and fire guns at the mining equipment of two companies.

Munkhbayar explained to me that the purpose of the 'operation to make a spark' was to provide a warning to the mining companies and the state, rather than to hurt anyone. They chose two companies to target: the Chinese-invested Puraam and the Canadian-funded Centerra Gold. These mines were selected for a number of reasons. First, for violations of the 'Law to prohibit mineral exploration and mining operations at headwaters of rivers, protected zones of the water reservoir and forest area'. Second, the Mongolian owners of these mines were high-ranking political authorities or their family members. Third, those mines were in historically and culturally sacred regions. Fourth, both of the mines had caused significant environmental damage. As the movement members discovered, Mongolian shareholders of the Puram mine included some high-ranking police and intelligence officers, while Batbold Sukhbaatar, the prime minister of Mongolia (2009–12), was the Mongolian shareholder of the Centerra Gold mine. According to the UMMRL, these mining companies committed crimes specified in Articles 202, 204, 205, 206, 207, 208 and 214 of the Criminal Law (Mongol Ulsyn Ikh Khural 2002; see also Simonov 2013). Both of the mines were located on the Noyon Mountain in Mandal, Selenge, in the north of Mongolia. The mountain is sacred, and it has dozens of burials from the time of the Xiongnu (3rd century BC–1st century AD). These were all considered excellent reasons for the movement leaders to choose these two mines and protest against them. Violations of criminal laws and the issue of the sacred historical site helped those movements to justify their actions against the two mines.

Immediately after the action, they organised a press conference to inform the public about the shooting. Further, they appealed to all Mongolians to support and join the movement. They explained that the action was to protect their homeland (*nutag*) and mother country (*ekh oron*) from mines that destroy the environment and the living conditions of ordinary people, which is a duty of every citizen of Mongolia as the constitution of Mongolia declares (Bügd Nairamdakh Mongol Ard Ulsyn Ardyn Ikh Khural 1992, Article 16, 17). They printed an appeal titled 'Let's save the Mongol hearth and our lives and livelihoods!' (*Mongol gal golomt, ami amidralaa avartsgaaya!*) and publicised it in the media.[1] The slogan and the above actions clearly depict the nationalist changes in emphasis and the narrative of the series of mobilisations led by Munkhbayar. Although legal action took place against the four members of the movement for shooting at the mining equipment, in April 2011, the court issued a judgment of innocence, and the incident of shooting was considered by the court as an act of desperate self-defence to stop the destruction caused by mining. In a media interview, the lawyer of the FN coalition of nationalist and environmentalist movements, Dashdemberel, stated that the court decision gave them hope and confidence to continue their fight (Erdeneburen 2011).

'The operation to make a spark' was one of two main actions of the FN. The FN leaders also actively engaged in promoting amendments to the 'Environmental Protection Law' (*Baigali orchnyg khamgaalakh tukhai khuuli*). In the 'Environmental Protection Law' Munkhbayar and Dashdembrel decided to lobby to add articles that would enable them to take legal action against the central government. This was something that the MNPC had promoted since 2005, but, as previously mentioned, they were unsuccessful in the 'Law to prohibit mineral exploration and mining operations at headwaters of rivers, protected zones of the water reservoir and forest area' in 2009. This time, based on their previous experience, the movement leaders took the decision to promote the amendment quietly and secretly. To achieve this, they decided to work with a different parliament member who was supportive during the process to promote the 'Law to prohibit mineral exploration and mining operations at headwaters of rivers, protected zones of the water reservoir and forest area'. Parliament member P. Altangerel introduced the amendment to parliament. In July 2010, the Parliament of Mongolia amended the 'Environmental Protection Law' (*Baigali orchnyg khamgaalakh tukhai khuuli*) (Mongol Ulsyn Ikh Khural 1995, Article 32), and so made it possible to sue the central government for the improper implementation of the law. Unsurprisingly, following this achievement, the next move of the FN was to take legal action against the government. In October 2010, leaders of the FN sued the government for improper implementation of the 'Law to prohibit mineral exploration and mining operations at headwaters of rivers, protected zones of the water reservoir and forest area'. Their appeal was moved down to the district court, but in October 2011, the Supreme Court found the government guilty and ordered it to enforce the law and to compensate mining companies affected by the law. The amendment in the 'Environmental Protection Law' and the activists' victory in the supreme court was an important turn in the conceptualisation of the state in Mongolia. The state officially became an institution that can be blamed and coerced into certain action, which was an instance and consequence of the de-deification of the state.

Both of the above victories of the FN coalition in the court, in April and October 2011, legitimised the actions of the movements and recognised failures on the part of the state (*töriin buruu*) and the illegality of the mining companies. This empowered these nationalist movements and their battles against mining companies and Mongolian authorities. The supreme court decision indicated that the movement was right, and also that it was capable of defeating both government and the mining companies, something that was previously far from obvious. These achievements fuelled the future actions of the FN coalition. The next campaign of the movement was an 'operation with horses' (*moritoi ajillagaa*). Munkhbayar and other leaders of the FN organised an event to bring about 100 horse riders from different parts of the country to protest in the central square in front of the state house (*töriin ordon*). Starting from around 14 April 2011, they gathered on the square. On horse carts, they brought eight *gers* (felt tents), which were set up on the square for almost two months to accommodate the protestors. The protest was effectively a depiction of nomadic herders' camps in the middle of the capital city, with horses tied up and dogs watching outside the *gers*. The *Tenger yazguur* (Heavenly lineage) society donated the *gers*. For about two months, the movement leaders requested a dialogue with the state authorities, but it was not taken up. At one point, they even went so far as to try to enter the residence of the president, prime minister and speaker in Ikh Tenger, in the Bogd Khan Mountain. On 3 June 2011, movement leaders shot five arrows at the windows of the state house. One of the arrows broke a window.

As became clear, it was expensive for the movement organisers to keep all of these men and horses in the city for two months. Munkhbayar later explained to me that the movement organisers had to pay for the herders' food to support their stay in Ulaanbaatar throughout the period. Munkhbayar said the operation with horses cost around 30 million MNT. Aside from some donations, to finance the operation Munkhbayar had to sell and pawn all of his possessions (Otgonsuren 2011). Later, Munkhbayar told me that he has spent almost all of the US$125,000 he received from the Goldman Prize to finance the mobilisations since 2007. That prize was equivalent to about 150 million MNT. He bought a two-bedroom flat in Ulaanbaatar for 47 million MNT, and a small second-hand Japanese four-wheel drive vehicle for 10 million MNT. With the rest of the money, he financed the movement, as it decided to stop receiving funding from TAF in 2007. He jokingly told me that the Goldman Prize funded the promotion of the 'Law to prohibit mineral exploration and mining operations at headwaters of rivers, protected zones of the water reservoir and forest area' and the activities to pressure the government to implement the law. By the time of the operation with horses, Munkhbayar was running out of money. To continue financing the campaign he had to sell the two-bedroom flat and bought a smaller flat. He used the money he made from the downsize to fund the horse operation. When he exhausted those funds, he then sold the little flat and used the cash to continue the protest.

On 29 June 2011, because it was expensive and difficult to control all of the horsemen in Ulaanbaatar for such a long period (and they had run out of money), they were forced to move on to the next operation called *nutag chölöölökh ajillagaa* (operation to liberate homeland). After two months, the number of riders had dropped to 33. Munkhbayar organised the remaining horsemen into two groups. One of the groups travelled to the north, while the other group went south. They visited every mine operating in the forbidden areas of rivers and forests. When they visited mining companies, they demanded that the companies write an official letter admitting that they had violated the law and agreeing to stop their extractions. The group that went to the north closed the operation of about 20 mines. Munkhbayar went to the south and closed only two mines. According to Munkhbayar, some high-ranking police officers contacted him and told him that government officials wanted to meet him in the local police station to negotiate. However, this turned out to be a trap. Munkhbayar and his team were captured and imprisoned for a few days.

All of the above events made Munkhbayar understand that engagement with the state is the most fruitful avenue to pursue in order to accomplish the mission of those movements. He also realised that one way to access the state was to influence as many political leaders and politicians as possible. The main difficulty was the political structure based on the political party system. He claimed that political parties have agendas to gain absolute political power and they tend to prioritise the party agenda and reputation above everything else. For this reason, he decided to attempt to unite independent election candidates with no political party affiliations. In June 2012, before the parliamentary election, he managed to meet about 50 people, all of whom could potentially run in the election independently and might support these mobilisations. It was challenging to unite different individual candidates with varied backgrounds, purposes and agendas. Ultimately, he managed to persuade 13 candidates and only a few days before the election officially announced the coalition in a press conference. Munkhbayar did not initially intend to stand in the election himself. However, other members in the alliance convinced him to stand, since he was the central person in establishment of the coalition. When I asked him why he did not consider standing in the election himself, he explained that organisation was more essential and useful for the movement to accomplish its mission than for him to become a member of parliament. Partly due to the short notice and poor management, this attempt at election did not succeed.

Green Terror?

On the morning of 16 September 2013, during a protest of a dozen environmentalist and nationalist protest movements at Chinggis Square (now Sükhbaatar Square), Munkhbayar and his colleagues of the FN coalition – armed with hunting rifles commonly used by herders in the Mongolian countryside – appeared near the entrance of the state house (*töriin ordon*). There was a gunshot as state security services (*toriin tusgai khamgaalaltyn gazar*), intelligence (*Tagnuulyn yerönkhii gazar*) and police stopped and arrested them. No one was hurt. Eleven men of the FN coalition, led by Munkhbayar – most of them dressed in Mongolian traditional *deel*, hat, boots and armed with rifles and hand grenades – were arrested. Later in the afternoon, some explosives were found in the bin outside the Central Tower on the Sükhbaatar Square, only a few hundred metres away from the state house. On the following day, more explosives were found near the building of the Ministry of Environment and Green Development (see also Campi 2013). Although the court found the incident to be a form of *zandalchlal* (terrorism) and a *töriin esreg gemt khereg* (crime against the state) (see Ul-Oldokh 2014), for these men it was a symbolic staged action to seek attention but not to inflict any destruction (see also Snow 2010).

The national and international media were filled with extensive debate discussing whether this was an act of green or eco-terrorism (see also Tolson 2014; Larson 2014; Snow 2014). According to Munkhbayar, some political leaders, intelligence agencies and police had specific intentions to depict the incidents in the media as terrorism. For example, Munkhbayar explains that one of the factors that supported the attempts to depict them as terrorists was the gunshot heard during the incident. He argues that the security who arrested them intentionally caused the gunshot; it was intended to suggest that it was a violent and dangerous incident. Until the trial confirmed that state security agents fired the gun, it was unclear who had fired. The gunshot left the public free to think that the movement members were terrorists. Keith Harmon Snow (2014), in his online article, complained that 'the court refused to question state security agents and refused to investigate who had fired the shot'. Moreover, he wrote about how national media accused and pictured the activists as terrorists:

> After the September 16 protest, the media accused the protestors of 'organising a public event without permission' a 'mass murder attempt' and even 'attempted genocide'. Mongolia's *National Overview* magazine, a copycat of *Time* (*Ündestnii Toim* in the Mongolian language rhymes with *Time*), featured Goldman prize winner Ts. Munkhbayar on the cover, an old Russian rifle in hand, under the headline: 'НОГООН ТЕРРОР' – GREEN TERROR. (Snow 2014)

For many Mongolians, the news was dreadful; the public was devastated. It was unusual for Mongolians to resist the state with rifles, hand grenades and explosives. Mongolia is known for and proud of its reputation of making a peaceful democratic transition without incidents of political violence. It was also culturally inappropriate and unusual to bring rifles against the state, as in the past Mongolians have effectively worshipped the state and treated it with high respect (see also Bumochir 2004; Humphrey and Hürelbaatar 2006; Dulam 2009). Munkhbayar told me that these are the very reasons which prompted them to bring rifles, grenades and explosives to the protest: this was the best way to attract attention. This was accomplished to a certain extent. The absence and unprecedentedness of terrorism in Mongolia led to attention via the sensationalised media coverage. Yet for these same reasons, many Mongolians condemned their attempt and sharply scolded them for using rifles, grenades and explosives. So, although the activists managed to garner much attention, much of it was focused not on the environmental problem but on the acts of *zandalchlal* (terrorism).

The primary court found five out of seven activists guilty and charged them with *zandalchlal* (Mongol Ulsyn Ikh Khural 2002, Article 177.2) and for obtaining and possessing guns and explosives (Mongol Ulsyn Ikh Khural 2002, Article 185.1 and 185.2). The primary court condemned to prison those five men for 20 to 21 years. When the court announced their sentences, the movement leader Munkhbayar stated that 'we are sentenced for our attempt to protect our home country' (Enkhdelger 2014). Many other national and international environmentalist and nationalist movements issued strong disagreements with the court decision. On 16 September 2013, a group of environmental organisations and activists called a press conference and produced a letter to parliament and the central government. Their position did not support the use of arms but blamed the government for not listening to them. In the letter, they stated that they found the incident to be an attempt to warn (Simonov 2013). On 8 November 2013, the Sosnovka Coalition of Environmental and Indigenous Civil Organizations of Siberia and Russian Fareast sent a letter to the Mongolian authorities supporting the law to protect rivers and forest areas and demanded justice in Munkhbayar's case.[2] On 20 November 2013, the Goldman Prize called for a fair and transparent trial for Munkhbayar. In the call posted on their website, they state, 'While the Goldman Prize does not in any way condone the use of weapons or violence, the Prize is calling on the Mongolian authorities to ensure a fair and transparent trial for Munkhbayar and we will be monitoring the situation as it develops'.[3] US-based activists from Inner Mongolia, China, at the Southern

Mongolian Human Rights Information Center stated that 'Munkhbayar's actions highlighted the desperation of helpless Mongolian pastoralists, who had no choice but to resort to an unconventional approach to defend their land, rights and way of life after exhausting all other means' (Tolson 2014). Snow (2014) wrote about this incident: 'Symbolically armed with hunting rifles and antiquated weapons, the most courageous leaders of the grassroots *Fire Nation* sought to draw attention to corruption and collusion between government and foreign mining corporations. They are fighting to save their culture and people and their very way of life'. There is a long list of supporters of Munkhbayar, highlighting the concern about Munkhbayar and Mongolia's environment, and demonstrating how much attention their action managed to attract from all around the world. All of these factors contributed to pressure the Mongolian government and the appellate court to reconsider Munkhbayar and his colleagues' case carefully. The appellate court found those men guilty, but this time for extortion (Mongol Ulsyn Ikh Khural 2002, Article 149.3), terrorism (*zandalchlal*) (Mongol Ulsyn Ikh Khural 2002, Article 177.1 and 177.2) and illegally obtaining and possessing guns and explosives (Mongol Ulsyn Ikh Khural 2002, Article 185.2). At this stage, the court did not find the case to be felonious and condemned them to prison for between one and ten years. They appealed against the appellate court decision and went to the supreme court. On 27 June 2014, the supreme court decided not to change the convictions of the appellate court.[4]

The incident of the armed protest happened on a day when an irregular spring session of the State Great Khural (Parliament) was summoned to amend some laws. Activists chose this day because the irregular session was planning to amend several laws, including the 'Law to prohibit mineral exploration and mining operations at headwaters of rivers, protected zones of the water reservoir and forest area'. According to Munkhbayar, the staged protest helped them to delay the amendment for a few more years. Unfortunately, two of the members passed away in prison due to health problems. In February 2015, when some of the key activists were in prison, parliament approved the regulations act that amended the law. The amendment changed the distance of mining operations from river shores. The distance had previously been 200 metres, but the regulations act changed it to 50 metres, which permitted most mining licences and operations. Munkhbayar and Dashdemberel both told me that the purpose of the law to protect river and forest areas was entirely diminished by the amendments.

Thoughts of the State in Prison

In his demonstration of notions of rights over land and the history of pastoralism, David Sneath explains that debates on land 'reflects the awareness that Mongolian culture, indeed the existence of Mongolia as a political entity is the product of a history of contested frames for rights over land. The modern Mongolian nation was formed in the course of a struggle waged by pastoral society and its elite to resist the loss of public Mongolian grazing land to agriculture' (Sneath 2001, 49) or to mining in the case of these movements. In the court trial Munkhbayar as a pastoralist deployed the same argument, as I will show in the following.

Some media sources claim that at the end of the trial, Munkhbayar said, 'Usgüi bol altaar yakh ve, Ulsgüi bol amiar yakh ve' (No need of gold if there is no water; No need of life if there is no country). When I asked Munkhbayar about this phrase, he told me that he had paraphrased a quote by Chinggis Khan, a quote he learned from historian B. Baljinnyam (see also Shiirev 2017). According to the historian, Chinggis Khan said, 'Usgüi bolvol uulyn chinee altaar yakh ve, Uls mongol min mökhvöl utsan chinee ulaan amiar yakh ve' (No value in a mountain of gold if there is no water; No value in the red thread of life if my country Mongolia collapses).[5] During the arrest, he and his colleagues could have been killed. They all were aware that they could lose their lives and they were prepared for such a possibility. With this quotation, he intended to send a message to Mongolians: the original interest of the nation preserved in the teaching of Chinggis Khan should not be ranked lower than the rights of individual business owners and private companies, and there are some Mongolians who will die to protect this interest.

Munkhbayar also claimed that some of the ideas in his message to protect the environment and the country have already been adopted and declared in the constitution of Mongolia. For example, the constitution starts with the following passages.

> We, the people of Mongolia:
> Strengthening the independence and sovereignty of the state,
> Cherishing human rights and freedoms, justice and national unity,
> Inheriting the traditions of national statehood, history and culture,
> Respecting the accomplishments of human civilisation,
> And aspiring toward the supreme objective of building a human, civil and democratic society in our homeland,

Hereby proclaim the Constitution of Mongolia.
(Bügd Nairamdakh Mongol Ard Ulsyn Ardyn Ihkh Khural 1992)

According to Munkhbayar, lines two and four declare the importance of what he calls the interest of the nation to be independent and sovereign, and the concern to preserve the 'traditions' of historical statehood. Moreover, Article 16.2 declares that citizens of Mongolia have 'the right to a healthy and safe environment, and to be protected [by the state] against environmental pollution and ecological imbalance' (Bügd Nairamdakh Mongol Ard Ulsyn Ardyn Ihkh Khural 1992, Article 16.2). For him, the well-being and security of the environment and people is a significant part in the concept of the sovereignty and national security. There is a salient way of thinking about politics in Mongolia that sees the environment, land, territory and sovereignty as analogous (see Chapter 2), which explains why the movement leaders consider the issue of the environmental damage to be a failure to protect the sovereignty of the nation; that is, the *uls ündestnii niitleg yazguur erkh ashig* (the original and common interest of the nation). Further, Article 17.2 declares that 'It is a sacred duty for every citizen . . . to protect nature and the environment' (Bügd Nairamdakh Mongol Ard Ulsyn Ardyn Ikh Khural 1992, Article 17.2). Quoting these articles, Munkhbayar claims that he had to perform his 'scared duty' as the constitution article encourages, while the state failed to provide its citizens with the right to live in a safe and healthy environment. In other words, Munkhbayar's argument was that the obligation of the state to protect the environment is not only something in the minds of nationalist activists, but is written into the very constitution. If the idea that 'the state exists for the benefit of the nation' is nationalist, as Barrington (2006, 21) considers, then such nationalism is legitimised in the constitution of Mongolia. According to Munkhbayar's experience, these articles in the constitution not only permitted him to be a nationalist, but also provided justification when they shot at mining equipment and forced some mining companies to cease their operations. The same legal condition also brought the supreme court to accuse the government of not enforcing the law to protect the environment. Munkhbayar told me that he had an important question: Couldn't the same legal condition justify their staged protest with rifles and grenades at the entrance of the state house? He thought that the last court decision that found him and his colleagues guilty was made to declare the supreme authority of the sovereign state, to demonstrate that the state is something that cannot be resisted by its people using arms. As a result, five members of the protest, including Munkhbayar, spent about two years in prison. As noted above, two of them died while in prison due to health issues. The remaining three members were released on 5 November 2015, as a consequence of the 'Law of Petition' (*Örshööl üzüülekh tukhai khuuli*), which released about 2,000 out of the total 6,000 prisoners in Mongolia (see also Ikon.mn 2015).

Munkhbayar said that he could never recognise the current state in Mongolia as a 'legitimate state' (*jinkhene tör*). For him, the state is something that is more than the presence of institutions and agents who fail to rule. According to his experience, although state institutions and agents are present, they fail in their duty of protection. He complained that every state is expected to enforce the law to protect its environment, land, territory and sovereignty, while the state failed to respond to their numerous efforts as they sought protection (see also Bebbington 2012b, 222). Such an expectation is valid according to how Lhamsuren (2006) explains the concept of the state as a ruling institution that Mongolians have historically experienced. In other words, the current state repeatedly failed to meet the expectation and to show people its presence through the protection of the environment which would benefit the nation. The absence of the state left no option for activists other than to perform the duty of the state. The state punished them for completing their 'sacred duty' (*ariun üüreg*) to protect. For these reasons, Munkhbayar refuses to recognise the current state as a 'legitimate state'.

Moreover, he claimed that such absence of the state and its failure to protect the environment, land, territory and sovereignty could potentially generate not just the resistance of environmental protest but also a *coup d'état* (*töriin ergelt*). He thinks that a *coup d'état* can be the last solution to bring radical changes in the political structure and to bring the 'genuine state' back. He feels that the state established by Chinggis Khan was an example of the 'genuine state'. In his idea to restore this, he envisions an indigenous political system, or a modern version of the political system Mongols had during the rule of the Mongol Empire (1206–1368). The system he recommended restores the historical arrangement of the military units of tens, hundreds and thousands, and small and great councils called *khuraldai* (or *kurultai*) that can be an alternative to parliament (see also Munkherdene 2018). Most importantly, he argues that at the top of the state there was not the rule of man; instead, there was the power of the *Mönkh Tenger* (Eternal Heaven), as depicted by Mongolian historical documents. Munkhbayar explained to me that the power of the *Mönkh Tenger* was an equivalent of the law of nature (*baigaliin khuuli*), which I will discuss in Chapter 6.

This argument about the absence of the state and the idea to restore the imperial system was what Munkhbayar developed when he was in prison. In other words, while the state showed him its presence by imprisoning him, he developed his idea about the absence of the state. Here, what interests me is the historical construction of the law of nature, the power of *Mönkh Tenger*, how Chinggis Khan acknowledged them and how all of these enabled Munkhbayar to argue about the absence of the 'legitimate state'. For Munkhbayar, his attempts to protect the environment, land and herders' way of life were justified not only by the court decisions mentioned above but also by historical constructions. In other words, he claims that his actions are protected and defended by what Chinggis Khan believed in, which means that the activists did not do anything wrong (and therefore did not deserve to receive prison sentences). The historically constructed justification suggests that the demonstrators naturally followed the teaching of Chinggis Khan, the ruler of the 'legitimate state', and tried to save water from gold extractions. To do this, they showed that they are ready to risk their lives for the well-being of the country.

After fighting against the gold mining companies for about three years, Munkbayar and most of the other protestors started to discuss the failures of the state (parliament, the government, the president and other state institutions and regulations) from around 2003. To find a solution to the failure of the state, activists not only resisted the state but attempted to address its shortcomings. Initially, activists unsuccessfully tried to lobby the state to make a political decision to ban mining in the river and forest areas. Consequently, in 2009 they successfully lobbied the 'Law to prohibit mineral exploration and mining operations at headwaters of rivers, protected zones of the water reservoir and forest area' (*Gol, mörnii usats büreldekh, usny san bükhii gazryn khamgaalaltyn büs, oin san bükhii gazart ashigt maltmal khaikh, ashiglakhyg khoriglokh tukhai*) (Mongol Ulsyn Ikh Khural 2009), known as the 'law with the long name'. The activists' fight against the state not only found a way to ban mining with the 'law with the long name' but also a means to prove the state's failure in the supreme court and thus pressure the state to implement the law. This successful experience fighting against the state motivated Munkhbayar to develop his thoughts of state by conceptually divorcing the actual state from what he calls the imagined 'legitimate state' (*jinkhene tör*) and by de-deifying the former and deifying the latter. In this sense, he fought not only against the 'legitimate state', but against some people in the actual state who had failed to make the state 'legitimate' (*jinkhene*).

Notes

1 For more information, see http://mongolianviews.blogspot.co.uk/2010/09/statement-from-civic-groups-on-recent.html; www.tsahimurtuu.mn/index.php/stories/2012-05-10-04-52-01/1934-archive-story-1758

2 For the full version of the letter, see www.transrivers.org/2014/1073/

3 For the full version of the call, see www.goldmanprize.org/blog/goldman-prize-calls-fair-and-transparent-trial-tsetsegee-munkhbayar/

4 For official documentation and court decision record reference, see www.shuukh.mn/eruudavah/1168/view

5 Many other movements acknowledge and popularise such phrases. For example, on 26 January 2015, the Bosoo Khökh Mongol (Standing Blue Mongol) movement, in collaboration with some other movements, organised a mass strike to resist a gold mining operation in the sacred Noyon Mountain, in the north of Mongolia. In the demonstration, the organisers printed and exposed the phrase *Minii gazar shoroonoos burkhan guisan ch büü ög* (Do not give away my land even if god asks for it) in huge letters: three metres tall on a 15-metre-long background (For photos, see http://mass.mn/n/27208).

6

An Original Environmentalist Society

Until 2008, the Ongi River Movement (ORM) and its successor the Mongolian Nature Protection Coalition (MNPC) were supported by local governments and donors. The friction generated in the collaboration of the river movements coalition with TAF not only divided the coalition but also transformed six of them (that is, the MNPC) into an alliance that became progressively radical nationalist.[1] Furthermore, they began to resist not only mining companies but also the state (see Chapters 4 and 5). They adopted more indigenous and historical, or 'traditional' (*ulamjlalt*), features of the discursive resources available in Mongolia and deployed them in their acts of resistance. Yet such tactics were present since the beginning of the movement. For example, in 2002, ORM members staged a protest involving hundreds of camels at Sükhbaatar Square, right in front of the state house (*töriin ordon*) (High 2017, 56). The camels symbolically represented herders and their livelihood in the southern *gobi* regions of Mongolia. The use of camels in the popular protest was a tactic to deploy depictions of 'traditional' (see also Humphrey 1992, 377; Kaplonski 2005, 166; Marsh 2009, 12; and Sneath 2010, 251) culture to the public as an inherent part of the movement. The *ger* protest described in Chapter 5 built on this earlier action. David Sneath (2010, 251) argues that 'notions of both tradition and collective identity have become potential resources, particularly for politicians, to mobilise public support'. For example, in his paper, Sneath mentions how Munkhbayar successfully mobilised the river movements using ideas of collective identity: *Ongi golynkhon* (People of the River Ongi) (Sneath 2010, 262). While Sneath addresses the importance of collective identity in the ORM, in this chapter I address how the notions of 'tradition' help the activists to mobilise and respond to accusations that Munkhbayar and his colleagues were extortionists.

The first part of the chapter describes Munkhbayar's 'return' to the countryside after his release from prison and his claims of *malchin* (herder/pastoralist) identity. The essentialist and romantic image of pastoralism in Mongolia is an iconic and important national symbol, and many often consider pastoralists as bearers of 'traditional' culture (see also Myadar 2011). I will show that Munkhbayar's 'return' to the countryside has symbolic resonance not just in his claim of *malchin* identity but also to construct an ideological claim that Mongols were the bearers of 'an original environmentalist society', which is the focus of the second part of this chapter. In the second part, I borrow Marshall Sahlins's (1972) memorable 'rehabilitation' of hunter-gatherer societies, in which he argues that such groups were the 'original affluent society in their capacity to match their restricted means to limited wants'. Similarly, to rephrase the ideology constructed by Munkhbayar, mobile pastoralists are the 'original environmentalist society', which forms the focus of the second part of the chapter. This original environmentalist identity of pastoralists is a claim that helps Munkhbayar to respond to the accusation that the environmental activists extort mining companies for the purpose of extracting money, rather than genuinely caring for the environment. Munkhbayar argues that pastoralists did not learn or adapt Western environmental concepts and practices to protect the environment. Instead, the protection of the environment is inherent to their everyday herding practices and way of life. The claim of the authentic care of original environmentalist pastoralists also helps Munkhbayar to respond to another accusation that blames pastoralists for environmental destruction, which I will account for in what follows.

The image of pastoralism as a symbolic resource construction of national identity and *nüüdliin soyol irgenshil* or 'nomadic-culture civilisation' arise in part as a reaction to derogatory discourses and historical depictions of the Mongols by their permanently settled neighbours and foreigners, who portrayed nomads as 'backward and uncultured barbarians' (Khazanov 1984, 1–15, 228–63; Khan 1996, 127–31; Humphrey and Sneath 1999, 1; Sneath 2007, 39–41; Bumochir and Chih-yu 2014, 417; Bayar 2014, 440–3). According to B. Tsetsentsolmon, the notion of *nüüdliin soyol irgenshil* emerged in opposition to negative presentations of 'uncivilised Mongolians' (Tsetsentsolmon 2014, 435). She further elaborates the history of the construction of *nüüdliin soyol irgenshil* in the Soviet and post-Soviet eras, mainly by leading Mongol scholars, from the early twentieth century to the present. Subsequently, the Mongolian concepts *nüüdelchin* for a nomad and *nüüdliin mal aj akhui* for pastoralism[2] have largely positive connotations,[3] often playing a crucial role in national pride and holding an honoured place in the construction of national identity.

Yet some recent considerations – climate change, the increasing desertification of pastureland in Mongolia and the need to conserve the environment – have led to a search for causes of land degradation. Specific forms of pastoralism, with different scales of mobility, were identified as among the leading human factors in Mongolia and Inner Mongolia, China.[4] Many scholars, local and international NGOs, stakeholders and policymakers in Mongolia and elsewhere bring up similar concerns regarding the unsustainability of pastoralism. Ian Hannam, an expert in international and national legal policy and institutional systems for natural resource management, advised Mongolia and many other Central Asian countries to develop new legislative and institutional structures. According to his advice, 'Mongolian rangeland is degraded because herders are unable to apply sustainable grazing practices. Mongolian grassland is not valued, so its regulation and management have been avoided in the past. Herders continue to graze their livestock on public land unrestrained, where there is high competition for good pasture. They use public pasture and water free of charge and without initiatives to protect and properly use it' (Hannam 2012, 418).[5] In Mongolia, many more contribute to this discussion, namely national and international scholars, consultants, NGOs, stakeholders and policymakers. Thus, this conservationist viewpoint on pastoralism constructed an image of a harmful and unsustainable practice with adverse effects on pasture and natural resources.[6] While such conservationist discourses problematise pastoralism, there is an alternative approach that considers the sustainability of pastoralism in Mongolia. This approach recognises that pastoralism involves complex reasoning, knowledge and techniques that prevent herders from overgrazing or causing damage to nature, which I will demonstrate in detail below. In essence, I will draw attention to the manner in which the above conservationist construction of unsustainable pastoralism is different from the local understanding in Mongolia, and contradicts how herders perceive and interpret pasture degradation as well as their well-being in the natural environment (see also Bumochir 2017).

Sneath (2003, 441) finds discussions of unsustainable pastoralism to be a 'Western interest in promoting Western conservationist ideology and establishing and expanding protected areas to harbour wildlife and biodiversity'. Similarly, Elliot Fratkin and Robin Mearns (2003) point out that those who employ the concept of sustainability 'often blame poor rural producers for what are perceived to be unsound practices including slash and burn cultivation, overgrazing, and deforestation for fuelwood' (Fratkin and Mearns 2003, 113). Following Sneath, Fratkin and Mearns, I focus on how the discourse of unsustainability contributes to accusations of blame upon the herders.

The accusation of the unsustainability of pastoralism conflicts with the existing local, indigenous and 'traditional' understandings of pastoralism as sustainable. In Munkhbayar's understanding, 'traditional' Mongolian cosmology and herding practice do not prioritise human rights and the individual's well-being, and do not pronounce human domination and triumphalism on earth (see also Stépanoff et al. 2017). Instead, the cosmology holds that every being on earth has interdependent *khüin kholboo* (umbilical connection) which depicts the mother-child relationship with the umbilical cord ruled by the law of nature (*baigaliin khuuli*). Munkhbayar affirms that human beings must follow the law of nature, rather than ignore it. As the movement leaders repeatedly pronounce, Mongol pastoralists embody and transmit this knowledge and practice (see also Upton 2010), which I will refer to as the indigenous environmentalist cosmology (see also Vitebsky 1995). Many scholars also put forward the same argument. For example, Caroline Upton writes that the 'ideas of reciprocal, respectful relations with nature were central to belief and practice' (2010, 309) in Mongolia. Referencing Caroline Humphrey et al. (1993), P. Tseren (1996), and A. Terenguto (2004), she further writes that 'in everyday life, these were enacted for example through cultural prohibitions on cutting live trees, digging the soil and overuse of resources, including allowing livestock to damage or overgraze the pasture, nomadism being intrinsically connected with these beliefs and practices' (Upton 2010, 309). In this way, the second section of the chapter shows that many Mongolians consider pastoralism to be a sophisticated, sustainable technology, and claim that pastoralists do not destroy but protect the environment. By developing new meanings and deployments of the term *pastoralism*, activists not only recreate the 'nomadic' and 'pastoral' national identities that they perceive as under threat, but also construct an indigenous environmentalist understanding of an original environmentalism.

A 'Return' to Pastoralism

A hush came across the courtroom as a guard declared, '*Shüügchid orj irlee bostsgoono uu!*' ('All rise! Judges are coming'). Two women and one man in long, black gowns entered the room; the one in the middle announced the court case and the names of the people involved. With a very formal tone and cold manner, one judge asked the litigant to stand up and identify himself and his address. Munkhbayar, in a Mongolian traditional dark green *deel* (Mongolian coat), a shirt with the same colour and fabric, hat and boots, stood up and announced his name and identified himself as a *malchin* (herder or pastoralist) from Saikhan-Ovoo *sum* (sub-province), Dundgovi *aimag* (province). His self-identification as *malchin* from the countryside reminded everyone that he is someone other than just an environmental and nationalist movement leader. In addition to being an activist, he is also a father, a husband, a national hero (or a 'terrorist', according to some people), and a *malchin*, as he presents himself in the courtroom. Of course, he must have a home somewhere, and most people would consider his home to be in Ulaanbaatar, the capital city of Mongolia, where most of the protest movements happen. But as a *malchin*, his home is almost unavoidably somewhere in the countryside. As he claimed in the courtroom, his home is in the Gobi Desert, where some of the largest mines operate and impact the environment.

While Munkhbayar and his colleagues were in prison, accused of extortion and terrorism, a new label became associated with Munkhbayar and his associates: 'green terrorist' rather than 'activist' (see Chapter 5). It is not uncommon for people in Mongolia to view environmental activism as an easy way to make money, through extortion, harassment and pretending to be a patriot, nationalist and protector of nature and the environment. Even Munkhbayar, as well as G. Dashdemberel, the lawyer who drafted the 'law with the long name', did not entirely reject such suspicions; they acknowledged the possibility that some people might extort money from mining companies in the name of the environment. In his court appearance, Munkhbayar's *malchin* identity partly had the effect of allaying such suspicions. Emphasising this identity was an attempt to prove that Munkhbayar and his associates were genuine and honest in their fight to protect the environment.

In November 2015, not long after the movement members were released from prison, an advert on social media invited the public to the trial of Munkhbayar and N. Gantulga, executive director of the ORM, which they hoped would clear their names (*ner töröö nekhemjilekh*) of extortion charges. On 23 November 2015, I attended the trial with my colleague Byambabaabar Ichinkhorloo, as Munkhbayar stated his case against the accusation that they extorted money from the AUM Gold mining company.[7] Sh. Myagmardorj, director of the AUM Gold, apologised. He explained that the accusation of extortion was a 'misunderstanding' (*buruu oilgoltsol*). He also admitted that for compensation, the company agreed to pay Munkhbayar MNT 80 million and to pay Gantulga MNT 50 million. It is rather implausible that the whole scenario – extortion charges, prosecution and imprisonment in addition to the criminal charges for *zandalchilal* (terrorism) – was just a 'misunderstanding'. The trial leads to another conclusion: perhaps the initial accusation was an act of politically organised repression (*khelmegdüülelt*). In the televised news interview at the end of the trial, Munkhbayar declared that the charge and accusation against him and his colleagues was 'political repression' (*uls töriin khelmegdüülelt*). Further, Munkhbayar was confident that the incident in question – including the extortion charge and his prison sentence for the so-called act of terrorism – will one day be recognised for what it was: a performance by protestors to grab attention for the sake of the environment and *not* terrorism. He believes that exculpation is only a matter of time.

After the court meeting, Byambabaatar and I talked to Munkhbayar to arrange a meeting for an interview. He told us that he had scheduled many meetings in the two days before his scheduled 'return' to the countryside to prepare for the winter before the weather gets too cold. It was the end of November 2015, and the beginning of the coldest winter of the past two decades. Munkhbayar asked us whether we could meet early in the morning. We agreed to meet at 6 a.m. on the day he was leaving Ulaanbaatar. This was our second meeting at the studio flat he was renting in a dormitory on the eastern edge of Ulaanbaatar. Until then, Byambabaatar and I did not know that Munkhbayar did not own a home in Ulaanbaatar. Nor was he based in Ulaanbaatar, which is the centre and the capital of the country. He leads a very different life than many might expect. When we visited, he was renting a tiny room with a toilet and bathroom in the corner, and simple cooking facilities on the floor in the opposite corner. When we arrived, Munkhbayar and his wife were already awake and had prepared tea. Two of his children were still sleeping on the floor. There was a small space left unoccupied, probably where Munkhbayar and his wife had been sleeping, and we sat there to start our conversation. We started by discussing his home, which has been constantly moving since 1996.

Before the ORM, he was a herder who studied agrotechnology in college (*technikum*) and later journalism (graduating in 2001). In 1996, Munkhbayar was elected as a member of Saikhan-Ovoo *sum* Citizens' Representative Meeting (*Irgediin tölöölögchdiin khural*) and lived in the *sum* administrative district centre until he resigned in 2003. In 2004, his family moved to Ulaanbaatar and started renting rooms in different flats (*karma öröö*, or 'pocket room', meaning a pocket in a flat or apartment) and dormitories until he received the Goldman Prize in 2007, which granted him US$125,000 (about 150 million MNT). As described in Chapter 5, with the grant he bought a two-bedroom flat in Ulaanbaatar and funded protests with the rest of the prize money. Later, in order to continue the protest, he sold his flat and bought a smaller one, before finally selling this to fund the movement. When he sold the small apartment, his entire family was left without a home in Ulaanbaatar, except the *ger* (felt tent) his two sons were living in with a few cattle just outside Ulaanbaatar.

In 2013, just before he received his prison sentence, the family decided to move back to Saikhan-Ovoo *sum* to continue their herding way of life once and for all. This 'return' was not just a 'return' to the *nutag* homeland in Mongolia, but to revitalise their herder identity. Unfortunately, the family had to stay in the countryside without Munkhbayar. The family fell apart and dispersed (*tarchikhsan*) between the countryside and city until the time of Munkhbayar's release. As he put it, his children were staying with different relatives and friends in Ulaanbaatar while trying to continue their studies and work. While he was in prison, he had to sell some of his cattle to pay the rents, fees and other living costs. I heard from different sources that G. Uyanga, an activist who was elected to parliament in 2012, whom I introduced in the preface, helped Munkhbayar's family while he was in prison. As soon as he was released from prison, the first thing he did was to find a place for his children who were in Ulaanbaatar: the small room we visited for our discussion. Our conversation was interrupted by a telephone call around 7 a.m. It was a journalist from Mass Media Group who hoped to interview Munkhbayar. His next appointment reminded us once again that Munkhbayar had a truly tight schedule before his departure from Ulaanbaatar to set up his *ger* in the winter pasture before it gets too cold. He is no longer, and perhaps was never, based in Ulaanbaatar. Certainly, he is now fully occupied as a herder whose home is in Saikhan-Ovoo *sum* in the *gobi* (Dundgovi). His 'return' to the countryside was not limited to him only; it also transformed the ORM: eight of the nine board members are now herders, all of whom permanently live in the *gobi*.[8] This 'return' to the countryside is a dramatic change compared to how they started: the ORM began with no herders, but with chairs of *sum* Citizens' Representative Meeting and governors, among others (see Chapter 3). Munkhbayar noticed the problem Caroline Upton (2012, 247) observes in her demonstration of the incorporation of Ulaanbaatar-based river movements and local herders. She claims that 'disengagement between local herders and UB based river movement activists pose an apparent threat to sustained grassroots activism'. For the same reason of disengagement, Munkhbayar wanted the movement to belong to herders, centred in the countryside, not in Ulaanbaatar. He clearly sees the symbolic usefulness of his 'return' to herding. He is a local celebrity and he knows that he has lots of local support.

Yet, during the years of active environmental protests, he developed alternative identities in addition to his *malchin* identity. On the Goldman Prize website, his biography summarised his trajectory as 'from herdsman to statesman'.[9] Similarly, his National Geographic Emerging Explorer profile depicts him as an 'emerging explorer' who grew up herding on the bank of the Ongi River and suffered from the consequences of the damage done to it.[10] In this way, his environmentalist identity is broadly known. Yet his herder identity had been playing an underlying role in explaining and narrating the success of his environmentalist and conservationist career. In Mongolia, the media and public rarely identify him as a herder but often depict him as a leader of various environmentalist movements and organisations of which he has been Chair, namely the Ongi River Movement (*Ongi golynkhon khödölgöön*), United Movement of Mongolian Rivers and Lakes (*Mongolyn gol nuuruudyn negdsen khödölgöön*), Mongolian Nature Protection Coalition (*Mongol nutag minu evsel*), and the Fire Nation (*Gal ündesten*) coalition. Although his *malchin* identity had long been silenced and overlooked, these herder roots were finally retrieved, revitalised and publicly announced when Munkhbayar declared in the court that he was a *malchin* from the *gobi*. The *gobi* is also where we had our next meeting.

In mid-December, Byambabaatar and I travelled 400 kilometres from Ulaanbaatar to meet Munkhbayar again. The winter ride took the whole day: we travelled in the snow, half on paved roads and half on dirt roads, from early morning to late at night. We arrived in the dark, around 9 p.m. The darkness made it challenging to locate Munkhbayar's newly erected *ger* in the winter camp. Mobile phone network coverage barely functions in the area, which made finding him even more difficult. The snow-covered dirt road showed no tracks or wheel-prints. We looked for his *ger* for about an hour. Finally, Munkhbayar decided to send a young herder – who was also a member of the ORM – to look for us with a torch in the darkness. After we spotted the flashing light, it took us another half an hour to reach his *ger* in the bushes nearby the frozen Ongi River. It was indeed a perfect place for a winter camp, located in the lowland within tall shrubs, well-hidden from wind and the coldness of the open steppe, and equally, from our earlier attempts to find it. As soon as we arrived, we noticed a few hundred sheep outside his *ger*. We were surprised to see the sheep; less than a month before, in Ulaanbaatar, he told us that he did not have any livestock except for a few cattle. Meeting us outside the *ger*, he smiled and enquired whether we had a pleasant trip to his place, and hoped that we did not have many difficulties or delays in finding his *ger*. Inside, it was warm and cosy, with a new floor covering and carpets. In the *khoimor* – the most respectful part of the *ger*, opposite the entrance – there were three orange chests. The first one had a portrait of Chinggis Khan on it; the middle one had two large books: one was a publication of Chinggis Khan's teachings, the other a volume of his decrees. The third chest had a certificate of the Goldman Environmental Prize. Everything else, inside the *ger* and outside, projects the image of a proper Mongol *ail* (family and household), especially when compared to his rented dormitory room, which we visited in Ulaanbaatar. We asked him whether he had managed to settle down in the winter camp and how the local herders in the area had welcomed him. Besides setting up his *ger* in the current location, he had already organised a board meeting of the ORM. An immediate outcome of the board meeting was the 200 sheep (and goats) outside his *ger*, donated from herders of five *sums* along the Ongi River to support the movement. In the board meeting, Munkhbayar shared his concern about donor organisations, advocacy and the necessity of self-funding to sustainably continue the movement (see Chapter 5). The board members decided to start a herd for the movement, which could serve the purpose of funding its activities. Munkhbayar shared with us his surprise: within a week there were donations of sheep from local herders of five *sums*. All of the five *sum* representatives who participated in the meeting decided to collect about 100 sheep from each *sum*. Most herders were happy to voluntarily donate one or two or more livestock to keep the movement running. In this way, the story of Munkhbayar and the ORM involves the construction of a grassroots movement, which started with local government employees and foreign donor funding and transformed into a herders' movement with an increasing level of attempts to self-fund.

Munkhbayar did not 'return' alone. He also brought the nation's most successful, powerful and globally known environmental movement with him. His 'return' challenges the environmentalist perspective on pastoralism's unsustainability and the harm it inflicts upon the environment, which I turn to below. Despite this confrontation, his 'return' to pastoralism is symbolically useful mainly because a mobile pastoralist identity is the best fit for him to employ as a grassroots environmental activist.[11] Furthermore, his return is useful to construct his concept of an original environmentalism and to argue that Mongol pastoralists are the original environmentalists, who have been preserving the environment in their way of living from generation to generation. Therefore, with his 'return' to the pastoralist way of life and claim to the nomadic pastoralist identity, he is not ending his protest but continuing it by appealing for an original environmentalist society.

An Original Environmentalism

In 2015, on a popular television programme, the director of the Gatsuurt company Chinbat Lhagva[12] (for more on Chinbat Lhagva, see Chapter 3) stated that pastoralism is more harmful to the environment than mining. His statement suggests that pastoralism is unsustainable, an argument that I presented in the introduction of this chapter. As an influential public figure in Mongolia, who grew up in a herding family, trained as a geologist and with mining experience, his statement was a powerful message to many Mongolians. Chinbat claims that herding takes a vast territory and the size of the degraded land is often much larger than areas damaged by mining operations. In many ways, this is Chinbat's response to herders' complaints against mining-induced environment damage.

In response to the notion of unsustainable pastoralism of this sort, Munkhbayar gave an account of why and how *malchin* identity is a genuinely conservationist one. For Munkhbayar, unsustainability has little to do with the nature of *mal aj akhui* (pastoralism). Instead, he argues that it is all about what the 'West' (*baruunykhan*) is *doing* to Mongolia. Munkhbayar continued that the West glorifies its civilisation as the most ideal and sophisticated in the world. In doing this, they also blemish (*gutaakh*) other cultures by criticising indigenous cultural principles (*gol amin yum*). In the case of Mongolian culture, its principle is *mal aj akhui*, and donor organisations propagate an understanding that *mal aj akhui* is harmful (*yavuurgüi*) and does not have a future. He explains that one should understand this in the framework of how the West labels other cultures as risky, vulnerable, uncertain and unsustainable. According to his view, if *mal aj akhui* is 'unsustainable' and harms the environment, this could only be true if it was something that Mongolian herders recently adopted. For him, any harm can only be a consequence of the transformation of Mongolian cosmology based on the worship of *Mönkh Tenger* (Eternal Heaven), and its replacement with other foreign views embodied in Tibetan Buddhism, Russian Communism and Western democracy and capitalism. To keep pastoralism sustainable, he argues, Mongolians need to reclaim its 'traditional' cosmology of *Mönkh Tenger* (Eternal Heaven).

Munkhbayar also talked about the foreignness of anthropogenic forces in Mongolia. According to his understanding, in sedentary culture the domination of man over nature establishes a shared sense that anything can be done for the sake of human beings, and jeopardises the existence of all other creatures and species on Earth,[13] while such an understanding is absent in the history and culture of the Mongols. Munkhbayar claims that today this ideology is coming to a dead end (*mukhardal*).[14] Contrary to 'Western' opinion, the cosmology of the Mongols does not make human beings central, does not urge one to dominate and conquer earth, and does not consider the earth to be granted solely for the human being (see also Stépanoff et al. 2017). Munkhbayar argues that the understanding of Mongolian pastoralists prevented them from inflicting anthropogenic forces on the environment and making pastoralism unsustainable.

To understand the current environmental problems and reconstruct authentic Mongolian pastoralism, Munkhbayar decided to reclaim what he considered the genuine Mongolian cosmology or beliefs (*itgel ünemshil*) (see also Sneath 2001, 45–6; Bruun 2006, 232; Marin 2010, 164; Upton 2010, 305). In his explanation, he repeatedly used three keywords: *khüi elgen sadan* (umbilical liver relatives), *baigaliin khuuli* (law of nature), and *Mönkh Tenger* (Eternal Heaven). The first one, *khüi elgen sadan* depicts everything on earth that is not man-made. He claimed that all things are connected by a familial relationship, just as an umbilical cord connects mother and child. Second, there is only one thing that is perfect in the world, and that one thing is the *baigaliin khuuli* (law of nature). Munkhbayar sometimes uses *Mönkh Tengeriin khuuli* (the law of the Eternal Heaven) to mean the law of nature. He believes that as nomads and pastoralists the Mongols have been learning, practising, testing, acknowledging and following this law throughout their history. This law creates an order and arranges the relationships of every human and nonhuman thing and species via umbilical connections (see also Sneath 2001, 45–7; Marin 2010, 164; Upton 2010, 308). Finally, *Mönkh Tenger* (Eternal Heaven) is the supreme deity of the Mongols, broadly found both in the folk religious practices of lay people and shamanism, and the imperial culture from the middle ages (Dulam 1997). Many contemporary Mongolians consider shamanism and belief in the *Mönkh Tenger* to be the authentic and original Mongolian religion in contrast to Buddhism (see also Heissig 1980, 6–7; Bulag 2004, 110). Since the

1990s, belief in *Mönkh Tenger* and shamanic practices has dramatically increased; *Mönkh Tenger* was re-established as the contesting national religion against Buddhism (Bumochir 2014, 478–81). As many other works reveal, the heavenly father *Mönkh Tenger* and the mother earth *Etügen* are the supreme deities Mongols worship (Dulam 1989; Heissig 1980, 47–9). It is these deities that Munkhbayar now chooses to believe in and employ in his environmentalist perspective. He thinks that all of the environmental problems caused by anthropogenic forces are the consequences of the failure to submit to the law of nature. According to this law, the earth is not granted solely for human beings but is shared with all of the other species and creatures. Emerging from his pastoralist background, he claims that herders (*malchin arduud*) have played a pivotal role in carrying and transmitting this folk knowledge and practice, which has ensured environmental sustainability for thousands of years. In this way, the 'traditional' Mongolian cosmology and sustainable practice of the pastoral way of life embodies the original environmentalism. In other words, his ideology posits that *malchin* (herdsman) is the embodiment of a complex of knowledge, practice, belief and ethics dedicated to the sustainable natural environment. His activism, built on this foundation, is not only intended to save the rivers and pastures but also to rehabilitate *Mongolian* environmentalism that can be transmitted to the next generation.

These are not the beliefs of Munkhbayar – and other activists – alone. Scholars and policymakers in Mongolia have expressed similar ideas. For example, in 2002, the Council for Sustainable Development of Mongolia (led by the prime minister), published *Tulkhtai khögjil – Mongolyn ireedüi* (*Sustainable Development – Mongolian Future*), an edited volume led by prominent Mongolian economist D. Dagvadorj (2002). The book argues that the fundamental feature of the nomadic way of life is to cohere with nature and the environment, and to secure their lives by protecting and restoring it. Similarly, other Mongolian scholars who publish on *nüüdliin soyol irgenshil* (nomadic culture and civilisation) and *nüüdeliin mal aj akhui* (mobile pastoralism) also reveal the same elaborations. For instance, S. Dulam writes in his *Mongol soyol irgenshliin utga tailal* (*Interpretation of Mongol Culture and Civilization*) that Mongols realise that nature cannot be produced again as it was; however, it can be preserved, and *nüüdeliin mal aj akhui* (mobile pastoralism) can preserve it in its original form (Dulam 2013, 29). Also, in the official state document called *Ündesnii ayulgüi baidlyn üzel barimtlal* (National security concept), passed in 2007, Article 47.1 was entirely dedicated to the preservation and enhancement of *nüüdliin soyol irgenshil* (Mongol Ulsyn Ikh Khural 2010).

I am not arguing that the whole concept is truly original, unique and entirely different from all other similar cosmologies in the rest of the world. Here, my intent is not to test whether Mongol herders maintain such environmental cosmology and do not overgraze their herds. What interests me is the difference that Munkhbayar creates and mobilises between the so-called Western environmentalist constructions of 'unsustainable pastoralism' and his construction of Mongolian environmentalist cosmology to imaginatively create the original environmentalist society.

Munkhbayar's indigenous environmentalist cosmology shares some essential similarities with Donna J. Haraway's (2003) account of 'companion species' and Anna Tsing's (2012, 152) concept of 'interspecies relationships' that are neglected in global capitalism. Tsing (2005, 249) notes that 'village elder, nature lover, and national activist perspectives are produced within different and somewhat autonomously formed understandings of nature'. What interests me here is not only the understanding of the relationship between man and nature but the ways such understandings serve the purpose of the indigenous environmental movement. For the environmental and nationalist movements, 'the autonomously formed understanding of nature' helps them to nationally and internationally publicise environmentalist discourses, respond to different accusations, and argue against the unsustainability of pastoralism. Unlike what we saw for Haraway and Tsing, in Mongolia the autonomous knowledge of man and nature is not a 'neglected perspective' but the knowledge that cannot be neglected. Thus, it becomes widely salient in political debate when it is effectively drawn upon and mobilised by the movement.[15] *Malchin* (that is, the pastoralist identity) – employed self-referentially by Munkhbayar – is inseparable from portrayals of imperial and heroic nomadic states and their civilisation, which is essential to the construction of the national identity, nationalism and independence of the Mongols (Bruun 2006, 227, 232; Upton 2010, 205). Reflecting on this, Munkhbayar's activism was a 'fight to remain with one's homeland' (*nutagtaigaa üldekhiin tölöökh temtsel*), and for him, it is a heroic expression of pastoralists' genuine care for their *nutag*. For this reason, Munkhbayar makes such knowledge the backbone of his nationalist sentiment, and a significant aspect in the making of an indigenous pastoralist environmentalist national identity that he seeks to further mobilise to achieve his political ends.

Notes

1 I do not mean that previously these movements were not nationalist at all, or that nationalism was absent. Also, when they become radically nationalist I do not mean that they stopped being environmentalist.

2 In my experience, there is an important difference between the Mongolian term *mal aj akhui* and its English form 'pastoralism', and also between the word *nüüdeliin* and its English form 'nomadic'. In Mongolian, these words do not have the connotations of backward, unsophisticated, unskilled or unsustainable, while their English forms do.

3 Not everyone supports the concept of 'nomadic civilisation'. Some philosophers, journalists and poets argue that civilisation is something to do with sedentary and urban culture (Bumochir and Chih-yu 2014, 417). For example, one of the latest encyclopaedias of Mongolian culture, *On the way Towards Civilisation: Almanah of Mongolian Culture* (*Irgenshliin zamd: Mongolyn soyolyn almanakh*), was written by well-known Mongolian poet and writer B. Tsenddoo (2015) and published by Nepko Publishing, which is owned by a famous journalist and public figure known as Baabar. Baabar is one of the pioneers who argued against the concept of nomadic civilisation from the early 1990s.

4 Environmentalist problematisation of pastoralism and 'green governmentality' in North China 'converted pastures to grasslands' (Yeh 2005; Kolås 2014), leaving many herders displaced and without herds. This turned many into ex-herders, but the former herders were also left with the anxiety of becoming ex-Mongols in Chinese urbanisation (see also Baranovitch 2016). N. Baranovitch writes about the 2011 protest in Inner Mongolia, sparked by the 10 May death of a Mongolian herder called Mergen, who was killed by a coal mining truck driven by a Han Chinese driver. He argues that the protest was not just about pasture degradation and mining destructions but also about the dying out of Mongolian culture and identity in China (2016, 228–30).

5 In the case of Inner Mongolia, Dee Mack Williams (2000) calls it Han Chinese 'scientific knowledge construction'. He writes that Han Chinese national and regional levels of government officials and scholars explicitly express that 'Mongols never learned to look beyond their sheep to the soil, so today they have no regard for the land that farmers have cherished' (2000, 508).

6 I must note that they are not saying that all pastoralists and their acts are universally harmful. Instead, the specific actions of some herders were found to be detrimental to the environment by some environmentalists.

7 I could not find any information about the company. From Munkhbayar and some other people who work in the Mongolian gold mining companies in Uyanga, where Aum Gold operates, I discovered that the company has investors from the Czech Republic.

8 In 2008, Upton observed the ORM to be more grassroots orientated than all other river movements in Mongolia. She writes that representatives, membership and awareness of the river movements were generally low among local herders, except in the immediate vicinity of the dried-up Ulaan Lake at the southern end of the Ongi River (Upton 2012, 244).

9 For the full profile, see www.goldmanprize.org/recipient/tsetsegee-munkhbayar/.

10 For the full profile, see www.nationalgeographic.com/explorers/bios/tsetsegee-munkhbayar/

11 S. Chuluun and G. Byambaragchaa (2014) talk about a different form of coexistence of herders in OT 'mine impact zones' in Khanbogd, Ömnögovi. Instead of resisting mines or attempting to stop them, they try to gain employment from the mining companies; they try to get a permanent salary to increase their financial resources.

12 For the full video, see www.youtube.com/watch?v=aVpBq1aM5K0

13 Indeed, many write how Europeans intended to master the earth. Consider, for instance, the Judaeo-Christian and Biblical traditions that can be found in Genesis, the first book of the Bible. Carl Hand and Kent Van Liere (1984), Ronald Shaiko (1987), Jeremy Cohen (1989) and many others show that the West has been developing the concept 'to master the earth' from its ancient religious teachings. William Leiss (1994) writes of the history of different concepts of mastery over nature and argues that another major facet of 'mastering of the earth', different from those based on biblical knowledge, was the age of the secular science and technology.

14 Another founding member of the FN coalition said something similar. In 2012, on a television talk show, Javkhlan Samand – a well-known singer who named the coalition of nationalist movements *Gal Ündesten* (Fire Nation) and who became a parliament member in 2016 – said

the following: 'When earth takes its saddle under its belly, and man eats each other, then livestock, herder and Mongol knowledge will save the world' (*Delkhii emeelee gedsendee avch, khün khünee barij idekh tsagt mal, malchin, mongol ukhaan gurav delkhiig avarch üldene*). By this he means that when the earth loses its ability to carry us like a horse on its saddle, and when man destroys each other and existing Western ideologies come to a dead end, then the mobile pastoral way of life and the knowledge of pastoralists will save the world. For the full version of the video, see www.youtube.com/watch?v=6QIiSMEcMHY

15 In contemporary Mongolia, many are rapidly abandoning the pastoral way of life and transitioning to urbanisation (Bruun and Narangoa 2006; Myadar 2011, 342). Some pastoralists evaluate their work as 'unskilled' and their subjectivities as 'uncultured', despite the widespread celebration of nomadic identities in nationalist discourse (Marzluf 2015). At the same time, some acknowledge that being a herder is not easy and does not assure a living, because harsh winters always risk wealth (see also Ericksen 2014). For these reasons, herders usually leave their children free to decide whether to stay and become a herder or do something else in urban areas, and most youngsters choose not to become herders (Ahearn and Bumochir 2016, 61–2). In addition to the herding communities, government officials also frequently devalue pastoral livelihoods and knowledge (89). Mongolian Prime Minister Enkhbayar Nambar expressed this view in 2001 when he called for the end of pastoralism and the movement of 90 per cent of the population to urban areas within 30 years (Endicott 2012). Enkhbayar supported his policy by questioning the viability of pastoralism after severe drought and winter weather killed the livestock of 12,000 families from 1999 to 2001 (Sternberg 2010). For many Mongolians, this is alarming because the number of those who carry the authentic nomadic and pastoralist national identity for the nation is dramatically decreasing and possibly ending. For example, in April 2016, national news claimed that every year Mongolia experiences a 2 per cent drop in the number of young herders aged 15 to 34 (For full video, see www.tv5.mn/index.php/society/1000-2016-04-01-11-01-59).

Conclusion

In April 2018, during the last stages of my fieldwork in Mongolia for this book, I met Dashdemberel Ganbold, the lawyer of the river movements and popular mobilisations who drafted the 'Law to prohibit mineral exploration and mining operations at headwaters of rivers, protected zones of the water reservoir and forest area', the 'law with the long name'. In the street, we briefly talked about his latest trials, the outcome of which had become less successful. He was surprised that he could no longer win court cases with the same legal conditions and arguments that he had used in the past. His success was affected by more than merely amendments made in different laws, such as the 2015 regulations act of the 'Law to prohibit mineral exploration and mining operations at headwaters of rivers, protected zones of the water reservoir and forest area' (see Chapter 5). In addition, activism designed to protect the environment – in general – no longer attracted as much public and political attention and support as it used to. He suggested then that it is not the laws and regulations that have decisive importance in the court; rather, it is the overall attitude of society at the particular time that matters the most.

Around the same time that I saw Dashdembrel, and throughout the time I was completing work on this book, I encountered dozens of social media posts concerning numerous cases of gold mining operations in the river and forest areas, and local protest movements against those mining operations from many different parts of Mongolia. One of those incidents was at the Zag River near my father's homeland (*nutag*), which I visited many times. In response to a Facebook post of a video titled 'Threatener of the life of the river Zag' (*Zagiin golyn ami nasand zanalkhiilegch*),[1] I noticed many people I knew from the Zag River and neighbouring regions – including my father – had liked and shared the post. There were also strong nationalist statements among the comments. I also noticed that the government of Mongolia was unresponsive and provided no reaction to protect those rivers. Instead, the central government was busy with the IMF bailout and dozens of strikes by government employees in the health care and education sectors.

Starting from around the end of 2013, the central government abandoned a strategy of concessions to the popular mobilisations against mining. Several factors helped to suppress popular mobilisations and allowed environmental problems to be ignored. One of these was the political labelling of the staged armed protest of September 2013 as terrorism and extortion (see Chapter 5). Such presentations created the impression that the popular mobilisations were not genuine. This also contributed to a prejudice that such movements have morphed into radical and dangerous extremist groups that should be avoided. Another factor was the economic crisis and neoliberal policies to assist the national economy.

May 19, 2015, was a historic day not only for Mongolia but also for the international community involved with developing one of the world's largest copper deposits, Oyu Tolgoi. After almost two years of ongoing disputes, the government of Mongolia and Rio Tinto signed a US$5.4 billion deal to expand the underground development of Oyu Tolgoi. Upon the signing, in the media the Prime Minister Saikhanbileg Chimed, who was committed to promoting the agreement, declared that 'Mongolia is back to business' (*Guardian* 2015, see Chapter 2). It can be assumed that the national government plan was to appeal further to foreign investors to prepare and secure its economy to pay over half a billion-dollar debt, starting from 2017 (Dulam 2015).

In August 2016, Finance Minister Choijilsuren Battogtokh announced that Mongolia is 'in a deep state of economic crisis' and 'the goal of the government is to avoid default'.[2] To help its economy, the government of Mongolia requested a rescue loan from the IMF (Hornby and Khan 2016). In May 2017 the IMF approved a US$5.5 billion bailout package for Mongolia (Edwards 2017). Mongolia's foreign debt – which started to increase dramatically around 2013 (see also Batsuuri 2015) – reached US$27.9 billion in March 2018[3] and the debt-to-GDP ratio reached about 90 per cent (Bauer et al. 2017, 1).

Another narrative emerged at this time that served to enforce the regime of the resource economy, which concerned China's 'debt trap' (*öriin urikh or öriin zanga*). In this way, failures in Mongolia's economy might make the country dependent not only on the IMF but also on China. According to some international media commentators, Mongolia became one of the victims of China's 'debt-trap diplomacy' (Fernholz 2018; Chellaney 2017). This news immediately alarmed many Mongolians and helped to enhance the anxieties surrounding political independence and the loss of land and territory. In this manner, the discourse of sovereign debt promotes the importance and legitimacy of the resource economy and undermines the environment to pay off the national debt.

This book narrates the political economy of resources, mining and environmental movements in Mongolia, which emerged after the collapse of socialism in 1990, when many politicians, economists and rulers of the nation-state established a free market economy with some urgency (Rossabi 2005; Munkh-Erdene 2012; Addleton 2013) to help combat the critical economic downturn. There was a significant focus on 'mining capitalism' in the emerging political and economic climate (Kirsch 2014). The operation of open-pit gold mines that used old Soviet technology (see Dear 2014; Bonilla 2016; High 2017) became one of the primary businesses and a vital financial resource for emerging Mongolian companies and the government. Mining helped to build the national economy. For this purpose, many rulers of Mongolia, including Ochirbat Punsalmaa – former miner, technocrat and the first president of Mongolia (1990–7) and the Democratic Party – were committed to the mining industry, market economy, neoliberalism and resource economy (see Chapter 1). Consequently, in 2012, the mining boom made Mongolia one of the fastest-growing economies, with a 17.3 per cent GDP growth rate (World Bank 2012). That same year, the massive gold and copper mine Oyu Tolgoi, which is a joint venture of Rio Tinto and the Mongolian government, made Mongolia the economy with the most investment (Riley 2012). Unfortunately, the so-called mining boom did not last long. Mongolia's GDP growth rate dramatically dropped and reached –1.4 per cent in 2016 (Trading Economics 2018). Thereafter, discourses of terrorism, economic crisis and sovereign debt helped to silence many actors in the state institutions in regard to mining-induced environmental damage, and endorsed the regime of the liberal economy of resources. However, matters of indigenous culture, history and environment in Mongolia remain, forming a compelling political resource of mobilisation that cannot be quickly suppressed or permanently terminated. Under the new government that took power in 2016 the environment, indigenous culture and history have gained prominence again.

Although mining successfully contributed to the Mongolian economy, its impact was also damaging; many nationalist protestors, local environmental movements, politicians, technocrats and scholars resisted and fought against political forces that promoted neoliberal policies. Since the 1990s, in parallel to the emergence of neoliberalism, the reconstruction of the 'traditional' culture, history, nomadism, pastoralism and national identity mainly reinforce ideas related to the protection of the environment, nature and lifestyles of local herders. As a result, nationalist environmental movements successfully brought problems related to the environment and local people in the areas deeply affected by mining to the attention of the national and international public, and to the consideration of the central government, parliament and president. In Mongolia, the simultaneous emergence of nationalism and statism and the circulation of neoliberal policies and capitalist market explains why popular mobilisations have emerged with so much power; these popular mobilisations were able to construct a regime that defined itself in opposition to the dominance of the resource economy.

The contemporary Mongolians presented in this book exist within a unique era: they are working to build the nation and to craft the state. In this politically prioritised and nationalised process to build a nation and to craft a sovereign state, the priorities of many modern Mongolians fall into two major spheres: one promotes the resource economy, global capitalist market, neoliberal policies and democracy; the other promotes history, tradition, environmentalism, naturalism and spiritualism. The so-called liberal reformers and nationalists or populists have ideological conflicts and disagreements and endlessly debate different issues. All of the chapters of this book show how those ideological conflicts, disagreements and debates in the process of crafting the state shape different practices and concepts, such as the national economy, mining, neoliberalism, mobilisation, nationalism, pastoralism, environmentalism, sovereignty and state. The process of building the nation continues, and the continuation of the process further generates and shapes policies and practices. In the context of the political economy of resource, mining and mobilisation in Mongolia, nation-building and state-crafting is the legitimate, most potent and resourceful political and discursive agenda.

The relationship between neoliberalism and the nation-state in Mongolia recalls Karl Polanyi's argument, which claims that it is impossible to dis-embed the economy from society and the state (Polanyi cited in Block 2001, xxvi–xxvii). Hanna Appel recently developed a similar argument. She states that 'there is no economy without the state' (Appel 2017, 301). Following Polanyi and Appel, as I mentioned earlier, throughout the book I consider neoliberalism as an ideology that can be employed for national purposes and used in the nation-building and state-crafting project as policy. In this case, I argue that there is not just an indigenous response and resistance to neoliberalism, as Maria Bargh

(2007) shows, but also the indigenous shaping of neoliberal policies and processes of neoliberalisation, by embracing it and then altering it to make it suitable to the indigenous, ethnic and national principles. Politicians, economists and nation-state rulers make the best use of neoliberal policies and others attempt to limit, control, transform or even break its principles. In other words, the use of neoliberal ideology in Mongolia is about embedding neoliberal policies in local or national contexts and making them their own.

As I mentioned in the Introduction, each of the chapters are intended to individually present the employment of different ideological positions to generate different discursive resources and contribute to discussions of different natures. Chapter 1 presented state rulers' employment of neoliberal ideology and contributed to the discussion about the anxiety of political independence and the reification of national economy. Chapter 2 presented nationalist technocrats' employment of Buddhist ideology regarding the economy and nationalist historical perspectives on the state protection of natural resources and contributed to the discussion of alternative understandings of the economy and 'resource nationalism'. In Chapter 3 I gave the narrations of those who own, manage and operate mining companies and who deploy neoliberal ideologies and also endorse a nationalistic identity of 'national wealth producer' (ündesnii bayalag büteegch). The findings of this chapter engage with discussions of the varying degrees of power of mining companies and deconstruct the homogenous transnational image of mining companies. In Chapter 4 I described how local government authorities started to endorse the local grassroots and civil society positions and established a right and a recognisable way to protest. The achievements and influence of popular mobilisations in this chapter suggest to us the need to reconsider conventional discussions on the power of local resistance movements, and the relationship of the local activist groups and international donors. In Chapter 5 I described how nationalist activists employed historically constructed statist concepts and showed how this provides a new perspective on the discussion of the deification and reification of the state and its power to control and protect the environment, natural resources and territory. Finally, in Chapter 6 I showed how activists employed nomadic herder and mobile pastoralist positions to construct an ideology of an ancient original environmentalist society which adds to discussions of environmentalism, nomadism and pastoralism. Besides the above individual interventions and contributions, all of the chapters should be seen as forming an overarching argument regarding the indigenisation and nationalist shaping of neoliberal policies and the capitalist market economy of resources within in the broad project to build the nation and to craft the state.

Notes

1 For the full video, see www.facebook.com/shine.suvd/videos/1728337473954305/?fb_dtsg_ag=AdxlHTc9YXAYJAsScOHtkyA8RsaG4mrF7EdogvrTa3D8g%3AAdx7XacDJBkizsVvG aLL0rMrXwzxJU5_5kZ_ODep9ccjdg
2 For the full version of his announcement on Bloomberg TV Mongolia, see www.youtube.com/watch?v=qo9Qb0LT3kc
3 For details, see www.ceicdata.com/en/indicator/mongolia/external-debt#.W5g70gqPBPM.email

References

Abrams, Philip. 1988. 'Notes on the Difficulty of Studying the State', *Journal of Historical Sociology* 1(1): 58–89.

Acheraïou, Amar. 2011. *Questioning Hybridity, Postcolonialism and Globalization*. Basingstoke: Palgrave Macmillan.

Addleton, Jonathan S. 2013. *Mongolia and the United States: A Diplomatic History*. Hong Kong: Hong Kong University Press.

Ahearn, Ariell and Bumochir, Dulam. 2016. 'Contradictions in Schooling Children among Mongolian Pastoralists', *Human Organization* 75(1): 87–96. https://doi.org/10.17730/0018-7259-75.1.87

Amarsanaa. S. and Bayarsaikhan, N. eds. 2007. *Töriig tsochooson ongiinkhon* (People of Ongi who alarmed the state). Ulaanbaatar: Od Press.

Anderson, Benedict. 1983. *Imagined Communities: Reflections on the Origin and Spread of Nationalism*. London and New York: Verso.

Andreucci, Diego. 2017. 'Resources, Regulation and the State: Struggles over Gas Extraction and Passive Revolution in Evo Morales's Bolivia', *Political Geography* 61(6): 170–80. http://dx.doi.org/10.1016/j.polgeo.2017.09.003

Andreucci, Diego and Radhuber, Isabella. 2017. 'Limits to "Counter-Neoliberal" Reform: Mining Expansion and the Marginalisation of Post-Extractivist Forces in Evo Morales's Bolivia', *Geoforum* 84(1): 280–91. https://doi.org/10.1016/j.geoforum.2015.09.002

Appel, Hannah. 2017. 'Toward an ethnography of the national economy', *Cultural Anthropology* 32(2): 294–322. https://doi.org/10.14506/ca32.2.09

Atwood, Christopher. 2004. *Encyclopedia of Mongolia and the Mongol Empire*. New York: Facts on File.

Auty, Richard. 1993. *Sustaining Development in Mineral Economies: The Resource Curse Thesis*. London: Routledge.

Badamsambuu, G. 2007. 'Irged khüchtei bol dülii niigem dünkhüü bodlogyg ch öörchilj chadna' [When citizens are powerful even a deaf society and dull policy can change]. In *Töriig tsochooson ongiinkhon* [People of Ongi who alarmed the state] edited by S. Amarsanaa and N. Bayarsaikhan, 126–8. Ulaanbaatar: Od Press.

Ballard, Chris and Banks, Glenn. 2003. 'Resource Wars: The Anthropology of Mining', *Annual Review of Anthropology* 32(1): 287–313. https://doi-org.libproxy.ucl.ac.uk/10.1146/annurev.anthro.32.061002.093116

Bank of Mongolia. 2017. 'Mongol Bankind tushaasan altny khemjee 10 khuviar össön baina' [Amount of gold sold to the Mongol Bank increased with 10%]. Last accessed 11 August 2019. www.mongolbank.mn/news.aspx?id=1711&tid=1

Banzragch, Ch. 2004. 'Altain kharuulyn tuhaid' [About the patrol of Altai]. In *Manjiin erkhsheeliin üyiin Mongol* [Mongolia in the times of the Manchu occupation], edited by J. Boldbaatar, 86–92. Ulaanbaatar: Orbis Publishing.

Baranovitch, Nimrod. 2016. 'The 2011 Protests in Inner Mongolia: An Ethno-environmental Perspective', *China Quarterly* 225: 214–33. https://doi.org/10.1017/S0305741015001642

Bargh, Maria. 2007. 'Introduction'. In *Resistance: An Indigenous Response to Neoliberalism*, edited by Maria Bargh, 1–25. Wellington: Huia Publishers.

Barrington, Lowell. 2006. 'Nationalism and Independence'. In *After Independence: Making and Protecting the Nation in Postcolonial and Postcommunist States*, edited by Lowell W. Barrington, 3–30. Ann Arbor: University of Michigan Press.

Batsaikhan, O. 2007. *Mongolyn tusgaar togtnol ba Khyatad, Oros, Mongol gurvan ulsyn 1915 ony Khyagtyn geree (1911–1916)* [Independence of Mongolia and 1915 Kyagta Agreement between China, Russia and Mongolia (1911–1916)]. Ulaanbaatar: Admon Press.

Batsaikhan, O. 2009. *Bogdo Jebtsundamba Khutukhtu, The Last King of Mongolia: Mongolia's National Revolution of 1911*. Ulaanbaatar: Admon Press.

Batsukh, J., and Chinzorig, O. 2012. 'Mongolyn avyaslag bisnesmenüüd—Lhagvyn Chinbat' [Talented Mongolian businessmen–Chinbat Lkhagva]. Last accessed 13 February 2019. www.wikimon.mn/content/8318.shtml

Batsuuri, Haltar. 2015. 'Original Sin: Is Mongolia Facing an External Debt Crisis?' *Northeast Asian Economic Review* 3(2): 3–15.

Battuvshin, B. 2007. 'Khödolgöönd zütgesen on jilüüd min' [My years working for the movement]. In *Töriig tsochooson ongiinkhon* [People of Ongi who alarmed the state] edited by S. Amarsanaa and N. Bayarsaikhan, 115–16. Ulaanbaatar: Od Press.

Bauer, Andrew et al. 2017. *Fiscal Sustainability in Mongolia*. Ulaanbaatar: Gerege Partners and Natural Resource Governance Institute. https://resourcegovernance.org/analysis-tools/publications/fiscal-sustainability-mongolia

Bayar, Nasan. 2014. 'A Discourse of Civilization/Culture and Nation/Ethnicity from the Perspective of Inner Mongolia, China', *Asian Ethnicity* 15(4): 439–57. https://doi.org/10.1080/14631369.2014.939329

Bear, Laura. 2015. *Navigating Austerity: Currents of Debt Along a South Asian River*. Stanford, CA: Stanford University Press.

Beaudoin, Gary. 2000. 'Gold Test on the Toson Terrace Placer, Zaamar Goldfield of Mongolia', *World Placer Journal* 1(1): 1–9.

Bebbington, Anthony et al. 2008. 'Mining and Social Movements: Struggles Over Livelihood and Rural Territorial Development in the Andes', *World Development* 36(12): 2888–2905. https://dx.doi.org/10.2139/ssrn.1265582

Bebbington, Anthony. 2012a. 'Extractive Industries, Socio-Environmental Conflicts and Political Economic Transformations in Andean America'. In *Social Conflict, Economic Development and the Extractive Industry: Evidence from South America*, edited by Anthony Bebbington, 3–29. London and New York: Routledge.

Bebbington, Anthony. 2012b. 'Conclusions'. In *Social Conflict, Economic Development and the Extractive Industry: Evidence from South America*, edited by Anthony Bebbington, 216–28. London and New York: Routledge.

Behrends, Andrea, Reyna, Stephen and Schlee, Günther eds. 2011. *Crude Domination: An Anthropology of Oil*. New York: Berghahn.

Benson, Peter and Kirsch, Stuart. 2010. 'Capitalism and the Politics of Resignation', *Current Anthropology* 51(4): 459–86. www.jstor.org/stable/10.1086/653091

Bilguun. 2015. '"Urt nertei" khuuliin khüreend litsenz ezemshigchdiin khüseltiig avch ekhlev' [Started receiving licence holders' requests in the range of the 'law with the long name']. Last accessed 16 March 2019. http://chuham.mn/index.php?newsid=324

Blaser, Mario and de la Cadena, Marisol. 2017. 'The Uncommons: An Introduction', *Anthropologica* 59(2): 185–93.

Block, Fred. 2001. 'Introduction'. In *The Great Transformation: The Political and Economic Origins of our Time*, by Karl Polanyi, xviii–xxxix. Boston: Beacon Press.

Boas, Taylor C. and Jordan Gans-Morse. 2009. 'Neoliberalism: From new Liberal Philosophy to Anti-Liberal Slogan', *Studies in Comparative International Development* 44(2): 137–61.

Bold, S. 2013. 'Alt khötölböriin khüreend 66.6 ton alt olborlojee' [66.6 tonnes of gold has been exploited in the range of the gold programme]. Last accessed 25 October 2019. www.mongolianminingjournal.com/content/51668.shtml

Bold-Erdene, S. 2013. 'Urt nertees nökhön tölbör nekhegsdiin daraalal nemegdseer' [Number of those demanding compensation from the long name is adding up]. Last accessed 5 November 2019. www.mongolianminingjournal.com/content/51934.shtml

Bollier, David, and Helfrich, Silke eds. 2012. *The Wealth of the Commons: A World beyond Market and State*. Amherst, MA: Levellers Press.

Bonilla, Lauren. 2016. 'Extractive Infrastructures: Social, Environmental, and Institutional Change in Resource-Rich Mongolia'. PhD diss., Worcester, MA: Clark University.

Bremmer, Ian and Johnston, Robert. 2009. 'The rise and fall of resource nationalism', *Survival* 51(2): 149–58. https://doi.org/10.1080/00396330902860884

Brown, Clair. 2015. 'Buddhist Economics: An Enlightened Approach to the Dismal Science', *Challenge* 58(1): 23–8. https://doi.org/10.1080/01603477.2015.990826

Brubaker, Rogers. 1999. 'The Manichean Myth: Rethinking the Distinction between "Civic" and "Ethnic" Nationalism'. In *Nation and National Identity: The European Experience in Perspective*, edited by Hanspeter Kriesi, Klaus Armingeon, Hannes Slegrist, and Andreas Wimmer, 55–73. Zurich: Rüegger.

Bruun, Ole and Odgaard, Ole. 1996. 'A Society and Economy in Transition'. In *Mongolia in Transition: Old Patterns and New Challenges*, edited by Ole Bruun and Ole Odgaard, 23–43. Richmond, Surrey: Curzon.

Bruun, Ole. 2006. *Precious Steppe: Mongolian Nomadic Pastoralists in Pursuit of the Market*. Oxford: Lexington Books.

Bruun, Ole and Narangoa, Li. eds. 2006. *Mongols from Country to City: Floating Boundaries, Pastoralism and City Life in the Mongol Lands*. Copenhagen: NIAS Press.

Bulag, Uradyn. 1998. *Nationalism and Hybridity in Mongolia*. Oxford: Clarendon Press.

Bulag, Uradyn. 2004. 'Inner Mongolia: The Dialectics of Colonization and Ethnicity Building'. In *Governing China's Multiethnic Frontiers*, edited by Morris Rossabi, 84–117. Seattle: University of Washington Press.

Bulag, Uradyn. 2012. 'Independence as Restoration: Chinese and Mongolian Declarations of Independence and the 1911 Revolutions', *Asia-Pacific Journal* 10(52/3): 1–15.

Bulag, Uradyn. 2017. 'A World Community of Neighbours in the Making: Resource Cosmopolitics and Mongolia's "Third Neighbour" Diplomacy'. In *The Art of Neighbouring: Making Relations Across China's Borders*, edited by Martin Saxer and Juan Zhang, 121–45. Amsterdam: Amsterdam University Press.

Bumochir, Dulam. 2004. 'Cult of the State: State in the Culture of the Mongols', *Acta Mongolica* 3(215): 103–20.

Bumochir, Dulam. 2014. 'Institutionalization of Mongolian Shamanism: From Primitivism to Civilization," *Asian Ethnicity* 15(4): 473–91. https://doi.org/10.1080/14631369.2014.939331

Bumochir, Dulam. 2017. 'Afterlife of Nomadism: Pastoralism, Environmentalism, Civilization and Identity in Mongolia and China.' In *Pastoralist Livelihoods in Asian Dry Lands: Environment, Governance and Risk*, edited by A. Ahearn, and T. Sternberg, 17–41. Cambridge: White Horse Press.

Bumochir, Dulam. 2018a. 'Mongolia'. In *Routledge Handbook of Civil Society in Asia*, edited by Akihiro Ogawa, 95–109. London and New York: Routledge.

Bumochir, Dulam. 2018b. 'Generating Capitalism for Independence in Mongolia', *Central Asian Survey* 37(3): 357–71. https://doi.org/10.1080/02634937.2018.1493429

Bumochir, Dulam and Chih-yu, Shih. 2014. 'Introduction', *Asian Ethnicity* 15(4): 417–21. https://doi.org/10.1080/14631369.2014.942055

Buyandelgeriyn, Manduhai. 2007. 'Dealing with Uncertainty: Shamans, Marginal Capitalism, and the Remaking of History in Postsocialist Mongolia'. *American Ethnologist* 34(1): 127–47.

Buyantogtokh, D. 2007. 'Sain khüniig sanan dursakhyn uchir minu' [My reason to remember the good person]. In *Töriig tsochooson ongiinkhon* [People of Ongi who alarmed the state] edited by S. Amarsanaa and N. Bayarsaikhan, 89–90. Ulaanbaatar: Od Press.

Bügüde nayiramdaqu mongγol arad ulus-un yeke qural [The Great Khural [Parliament] of the Mongolian People's Republic]. 1940. 'Bügüde nayiramdaqu mongγol arad ulus-un ündüsün qauli' [Constitution of the People's Republic of Mongolia]. Assent June 30.

Bügd Nairamdakh Mongol Ard Ulsyn Ardyn Ikh Khural [The Great Khural [Parliament] of the Mongolian People's Republic]. 1960. 'Bügd Nairamdakh Mongol Ard Ulsyn Ündsen Khuuli' [Constitution of the People's Republic of Mongolia]. Assent July 6.

Bügd Nairamdakh Mongol Ard Ulsyn Ardyn Ikh Khural [The Great Khural [Parliament] of the Mongolian People's Republic]. 1992. 'Mongol Ulsyn Ündsen Khuuli' [Constitution of Mongolia]. Assent January 13. www.legalinfo.mn/law/details/367

Bügd Nairamdakh Mongol Ard Ulsyn Ardyn Ikh Khural [The Great Khural [Parliament] of the Mongolian People's Republic]. 1988. 'Mongol ulsyn khuuli: Gazryn khevliin tukhai' [Law of Subsoil]. Assent November 29. www.legalinfo.mn/law/details/218

Byambajav, Dalaibuyan. 2015. 'The River Movements' Struggle in Mongolia', *Social Movement Studies* 14(1): 92–7.

Çalışkan, Koray and Callon, Michel. 2009. 'Economization, Part 1: Shifting Attention from the Economy Towards Processes of Economization', *Economy and Society* 38(3): 369–98. https://doi.org/10.1080/03085140903020580

Callon, Michel. 1998. 'The Embeddedness of Economic Markets in Economics'. In *The Laws of the Markets*, edited by M. Callon, 1–57. Oxford: Blackwell.

Campi, Alicia and Baasan, Ragchaa. 2009. *The Impact of China and Russia on United States–Mongolian Political Relations in the Twentieth Century*. Lewiston, NY: Edwin Mellen Press.

Campi, Alicia. 2013. 'Mongolian "Eco-Terrorists" Attack Ulaanbaatar to Protest Looser Mining Laws'. Last accessed 30 September 2019. https://jamestown.org/mongolian-eco-terrorists-attack-ulaanbaatar-to-protest-looser-mining-laws/

Cashell, Lee. 2015. Mongolia hopes for a phase of economic stability. Last accessed 11 August 2019. www.apip.com/news/apip-in-the-news/mongolia-economic-stability

Chellaney, Brahma. 2017. 'Sri Lanka the latest victim of China's debt-trap diplomacy'. Last accessed 24 December 2017. www.atimes.com/article/sri-lanka-latest-victim-chinas-debt-trap-diplomacy/

Childs, John R. 2016. 'Geography and Resource Nationalism: A Critical Review and Reframing', *The Extractive Industries and Society* 3(2): 539–46.

Chuluunbat, Narantuya and Empson, Rebecca. 2018. 'Networks and the Negotiation of Risk: Making Business Deals and People among Mongolian Small and Medium Businesses', *Central Asian Survey* 37(3): 419–37. https://doi.org/10.1080/02634937.2018.1450221

Chuluundorj, Khashculuun and Danzanbaljir, Enkhjargal. 2014. 'Financing Mongolia's Mineral Growth', *Inner Asia*. 16(2): 275–300.

Chuluun, S. and Byambaragchaa, G. 2014. 'Satellite Nomads: Pastoralists' Tactics in the Mining Region of Mongolia', *Inner Asia* 16(2): 409–26.

Click, Reid W. and Weiner, Robert J. 2010. 'Resource Nationalism Meets the Market: Political Risk and the Value of Petroleum Reserves', *Journal of International Business Studies* 41(5): 783–803.

Cohen, Jeremy. 1989. *'Be Fertile and Increase, Fill the Earth and Master It': The Ancient and Medieval Career of a Biblical Text*. Ithaca, NY: Cornell University Press.

Collier, Paul. 2007. *Bottom Billion: Why the Poorest Countries are Failing and What Can Be Done About It*. Oxford: Oxford University Press.

Collier, Stephen. 2009. 'Topologies of Power: Foucault's Analysis of Political Government beyond "Governmentality"'. *Theory, Culture and Society* 26(6): 78–108.

Cordero, Raúl et al. 2005. 'Economic Growth or Environmental Protection? The False Dilemma of the Latin-American Countries', *Environmental Science and Policy* 8(4): 392–8.

Coronil, Fernando. 1997. *The Magical State: Nature, Money, and Modernity in Venezuela*. Chicago: University of Chicago Press.

Crossley, Nick. 2002. *Making Sense of Social Movements*. Buckingham: Open University Press.

Dagvadorj, D. 2002. *Tulkhtai khögjil: Mongolyn ireedüi* [Sustainable Development: Mongolian Future]. Ulaanbaatar: Council for Sustainable Development of Mongolia.

Daly, Herman. 1996. *Beyond Growth: The Economics of Sustainable Development*. Boston: Beacon Press.

Dashnyam, Gongor. 2014. 'Galdan Boshgot Sudlal' [The Study of Galdan Boshgot]. In *Galdan Boshgot Sudlal* [The Study of Galdan Boshgot], edited by E. Jigmeddorj and Na. Sukhbaatar, 225–31. Ulaanbaatar: Soyombo Printing.

Dear, Devon. 2014. 'Marginal Revolutions: Economies and Economic Knowledge between Qing China, Russia, and Mongolia, 1860–1911'. PhD diss., Harvard University.

Dierkes, Julian. 2016. 'Resource Nationalism?' Last accessed 31 March 2019. http://blogs.ubc.ca/mongolia/2016/resource-nationalism/

Dierkes, Julian and Jargalsaikhan, Mendee. 2017. 'Election 2017: Making Mongolian Great Again?' Last accessed 20 June 2019. https://thediplomat.com/2017/06/election-2017-making-mongolia-great-again/

Domjan, Paul and Stone, Matt. 2010. 'A Comparative Study of Resource Nationalism in Russia and Kazakhstan 2004–2008', *Europe–Asia Studies* 62(1): 35–62. https://doi.org/10.1080/09668130903385374

Dulam, Sendenjav. 1989. *Mongol domog züin dür* [Figures in Mongolian Mythology]. Ulaanbaatar: Ulsyn khevleliin gazar.

Dulam, Sendenjav. 1997. 'Mönkh tengeriin takhilgyn belgedel' [Symbolism in the Worship of the Eternal Heaven], *Erdem shinjilgeeni bichig* 116(8): 75–94.

Dulam, Sendenjav. 2013. *Mongol Soyol Irgenshliin Utga Tailal* [Meaning and Interpretation of Mongol Culture]. Ulaanbaatar: Bit Press.

Dulam, Bumochir. 2006. 'Respect and Power without Resistance: Investigations of Interpersonal Relations among the Deed Mongols'. PhD diss., University of Cambridge.

Dulam, Bumochir. 2009. 'Respect for the State: Constructing the State on Present Humiliation and Past Pride', *Inner Asia* 11(2): 259–81. https://doi.org/10.1163/000000009793066550

Dulam, Bumochir. 2015. 'Mongolia's "strategic mine" and the conflict between civil society, national government, and foreign investors'. Last accessed 30 June 2019. http://blogs. ucl.ac.uk/mongolian-economy/2015/06/30/mongolias-strategic-mine-and-the-conflict-between-civil-society-national-government-and-foreign-investors/

Economist. 1977. 'The Dutch Disease', *The Economist*. 26 November. 82–3.

Edwards, Gemma. 2008. 'The "Lifeworld" as a Resource for Social Movement Participation and the Consequences of its Colonization', *Sociology* 42(2): 299–316.

Edwards, Terrence. 2017. 'IMF approves $5.5 billion bailout package for Mongolia', *Reuters*, 25 May. https://uk.reuters.com/article/uk-mongolia-imf-idUKKBN18L0AC

Ekins, Paul. 2002. *Economic Growth and Environmental Sustainability: The Prospects for Green Growth*. London and New York: Routledge.

Els, Frik. 2012. 'Mongolia's champion of resource nationalism is going to jail for four years'. Last accessed 19 August 2019. www.mining.com/mongolias-champion-of-resource-nationalism-going-to-jail-for-graft-13500/

Empson, Rebecca and Webb, Tristan. 2014. 'Whose Land is it Anyway? Balancing the Expectations and Demands of Different Trusting Partnerships in Mongolia', *Inner Asia* 16(2): 231–51. doi: 10.1163/22105018-12340017

Endicott, Elizabeth. 2012. *A History of Land Use in Mongolia: The Thirteenth Century to the Present*. New York: Palgrave Macmillan.

Enkhdelger, L. 2014. 'Ts. Munkhbayar nart üzüülsen töriin tömör nüür' [Iron face of the state shown to Ts. Munkhbatar and others]. Last accessed 22 January 2019. http://old.eagle. mn/content/read/11745.htm

Erdeneburen. O. 2011. 'Gal Ündesten kholboony udirdakh zövlöliin gishüün Ts. Munkhbayar: Mongolyn baigali orchin ezengüi bolson' [Board member of the 'Fire Nation' coalition Ts. Munkhbayar: Environment of Mongolia became ownerless]. Last accessed 12 April 2019. http://news.gogo.mn/r/86024

Ericksen, Annika. 2014. 'Depend on Each Other and Don't Just Sit: The Socialist Legacy, Responsibility, and Winter Risk among Mongolian Herders', *Human Organization* 73(1) 38–49. https://doi.org/10.17730/humo.73.1.e218g65507n665u0

Eriksen, Thomas Hylland, Laidlaw, James, Mair, Jonathan, Martin, Keir and Venkatesan, Soumhya. 2015. 'The Concept of Neoliberalism has Become an Obstacle to the Anthropological Understanding of the Twenty-First Century', *Journal of the Royal Anthropological Institute* 21(4): 911–23. doi:https://doi.org/10.1111/1467-9655.12294

Escobar, Arturo. 2008. *Territories of Difference: Place, Movements, Life, redes*. Durham, NC: Duke University Press.

Fratkin, Elliot and Mearns, Robin. 2003. 'Sustainability and Pastoral Livelihoods: Lessons from East African Maasai and Mongolia', *Human Organization* 62(2): 112–22. https://doi.org/10.17730/humo.62.2.am1qpp36eqgxh3h1

Ferguson, James. 2006. *Global Shadows: Africa in the Neoliberal World Order*. Durham, NC: Duke University Press.

Fernholz, Tim. 2018. 'Eight countries in danger of falling into China's "debt trap"'. Last accessed 7 March 2019. https://qz.com/1223768/china-debt-trap-these-eight-countries-are-in-danger-of-debt-overloads-from-chinas-belt-and-road-plans/

Ganbold, Misheelt and Ali, Saleem H. 2017. 'The Peril and Promise of Resource Nationalism: A Case Analysis of Mongolia's Mining Development', *Resources Policy* 53(C): 1–11. https://doi.org/10.1016/j.resourpol.2017.05.006

Gankhuyag, D. 2007. *'Mongolyn ündesnii üzel mandtugai!'* [Long Life of Mongolian Nationalism!]. Last accessed 17 April 2019. www.tsahimurtuu.mn/index.php/stories/2012-05-10-04-48-56/1393-archive-story-1188

Gardner, Katy. 2012. *Discordant Development: Global Capitalism and the Struggle for Connection in Bangladesh*. London: Pluto Press.

Genota, Jeff. 2017. 'Whatever Mongolia's Presidential Results, Populism Will be in Check'. Last accessed 11 August 2019. www.cogitasia.com/whatever-mongolias-presidential-results-populism-will-be-in-check/

Gilberthorpe, Emma and Banks, Glenn. 2012. 'Development on Whose Terms?: CSR Discourse and Social Realities in Papua New Guinea's Extractive Industries Sector', *Resources Policy* 37(2): 185–93. https://doi.org/10.1016/j.resourpol.2011.09.005

Gilberthorpe, Emma and Rajak, Dinah. 2016. 'The Anthropology of Extraction: Critical Perspectives on the Resource Curse', *Journal of Development Studies* 53(2): 186–204. https://doi.org/10.1080/00220388.2016.1160064

Godoy, Ricardo. 1985. 'Mining: Anthropological Perspectives', *American Review of Anthropology* 14(1): 199–217. https://doi.org/10.1146/annurev.an.14.100185.001215

Goldin, Ian, and Winters, L. Alan. 1995. 'Economic Policies for Sustainable Development'. In *The Economics of Sustainable Development*, ed. Ian Goldin and L. Alan Winters, 1–16. Cambridge: Cambridge University Press. https://doi.org/10.1017/CBO9780511751905.002

Golub, Alex. 2014. *Leviathans at the Gold Mine: Creating Indigenous and Corporate Actors in Papua New Guinea*. Durham, NC: Duke University Press.

Graeber, David. 2011. *Debt: The First 5,000 Years*. New York: Melville House.

Gramsci, Antonio. 2000. *The Gramsci Reader: Selected Writings 1916–1935*. New York: New York University Press.

The Guardian. 2015. 'Rio Tinto and Mongolia sign multibillion dollar deal on mine expansion'. 19 May. Last accessed 11 August 2019. www.theguardian.com/global/2015/may/19/rio-tinto-and-mongolia-sign-multibillion-dollar-deal-on-mine-expansion

Gylfason, Thorvaldur, Herbertsson, Thryggvi Thor, and Zoega, Gylfi. 1999. 'A Mixed Blessing: Natural Resources and Economic Growth', *Macroeconomic Dynamics* 3(2): 204–25.

Habermas, Jürgen. 1987. *The Theory of Communicative Action*, vol. 2: *System and Lifeworld*. Cambridge: Polity Press.

Hand, Carl and Van Liere, Kent. 1984. 'Religion, Mastery-Over-Nature, and Environmental Concern', *Social Forces* 63(2) 555–70. https://doi.org/10.1093/sf/63.2.555

Hannam, Ian. 2012. 'International Perspectives on Legislative and Administrative Reforms as an Aid to Better Land Stewardship.' In *Rangeland Stewardship in Central Asia: Balancing Improved Livelihoods, Biodiversity Conservation and Land Protection*, edited by Victor Squires, 407–29. Heidelberg, New York and London: Springer. https://doi.org/10.1007/978-94-007-5367-9

Haraway, Donna J. 2003. *Companion Species Manifesto: Dogs, People, and Significant Otherness*. Chicago: Prickly Paradigm Press.

Harvey, David. 2005. *A Brief History of Neoliberalism*. Oxford: Oxford University Press.

Harvey, David. 2010. *The Enigma of Capital and the Crisis of Capitalism*. Oxford: Oxford University Press.

Hatcher, Pascale. 2014. *Regimes of Risk: The World Bank and the Transformation of Mining in Asia*. Basingstoke: Palgrave Macmillan.

Heissig, Walter. 1980. *The Religions of Mongolia*. Berkeley: University of California Press.

High, Mette and Schlesinger, Jonathan. 2010. 'Rulers and Rascals: The Politics of Gold in Mongolian Qing history,' *Central Asian Survey* 29(3): 289–304. https://doi.org/10.1080/02634937.2010.518008

High, Mette. 2012. 'Cultural Legends in Illegality: Living Outside the Law in the Mongolian Gold Mines'. In *Change in Democratic Mongolia: Social Relations, Health, Mobile Pastoralism, and Mining*, edited by Julian Dierkes, 249–71. Leiden: Brill.

High, Mette M. 2017. *Fear and Fortune: Spirit Worlds and Emerging Economies in the Mongolian Gold Rush*. Ithaca, NY: Cornell University Press.

Hilson, Gavin, and Laing, Tim. 2016. 'Guyana Gold: A Unique Resource Curse?' *Journal of Development Studies* 53(2): 229–48. https://doi.org/10.1080/00220388.2016.1160066

Hornby, Lucy and Khan, Mehreen. 2016. 'Mongolia asks IMF a rescue loan.' Last accessed 30 September 2019. www.ft.com/content/de5ab480-8713-11e6-a75a-0c4dce033ade

Howitt, Richard, Connell, John, and Hirsch, Philip, eds. 1996. *Resources, Nations, and Indigenous Peoples: Case Studies from Australasia, Melanesia, and Southeast Asia*. Melbourne: Oxford University Press.

Humphrey, Caroline, Mongush, Marina and Telengid, B. 1993. 'Attitudes to Nature in Mongolia and Tuva: A Preliminary Report', *Nomadic Peoples* 33(2) 51–61.

Humphrey, Caroline. 1992. 'The Moral Authority of the Past in Post-socialist Mongolia', *Religion, State and Society* 20(3–4): 375–89.

Humphrey, Caroline and Sneath, David. 1999. *The End of Nomadism? Society, State and the Environment in Inner Asia*. Durham, NC: Duke University Press.

Humphrey, Caroline. 2002. *The Unmaking of Soviet Life: Everyday Economies After Socialism*. Ithaca, NY: Cornell University Press.

Humphrey, Caroline and Hürelbaatar A. 2006. 'The Term Törü in Mongolian History'. In *Imperial Statecraft: Political Forms and Techniques of Governance in Inner Asia, Sixth-Twentieth Centuries*, edited by David Sneath, 265–94. Bellingham: WA Centre for East Asian Studies, Western Washington University Press.

Ichinkhorloo, Byambabaatar. 2018. 'Collaboration for survival in the age of the market: diverse economic practices in postsocialist Mongolia', *Central Asian Survey* 37(3): 387–404. https://doi.org/10.1080/02634937.2018.1501347

Idemudia, Uwafiokun. 2010. 'Corporate Social Responsibility and the Rentier Nigerian State: Rethinking the Role of Government and the Possibility of Corporate Social Responsibility in the Niger Delta', *Canadian Journal of Development Studies* 30(1): 131–51. https://doi.org/10.1080/02255189.2010.9669285

Il Tod. 2012. '"Gol, mörnii ursats büreldekh, usny san bükhii gazryn khamgaalaltyn büs, oin san bükhii gazart ashigt maltmal khaikh, ashiglakhyg khoriglokh tukhai" khuuliin kheregjiltiin tukhai' [About the implementation of the 'Law to prohibit mineral exploration and mining operations at headwaters of rivers, protected zones of the water reservoir and forest area']. Last accessed 1 March. www.iltod.gov.mn/?p=2513.

Ikon.mn 2015. '"Gal Ündesten" kholboony tergüün Ts. Munkhbayar sullagdlaa'. ['Fire Nation' coalition leader Ts. Munkhbayar is released]. Last accessed 5 November 2019. https://ikon.mn/n/lix

Jackson, Sara. 2015. 'Imagining the Mineral Nation: Contested Nation-Building in Mongolia', *Nationalities Papers: The Journal of Nationalism and Ethnicity* 43(3): 1–20. https://doi.org/10.1080/00905992.2014.969692

Jackson, Sara, and Dear, Devon. 2016. 'Resource Extraction and National Anxieties: China's Economic Presence in Mongolia', *Eurasian Geography and Economics* 57 (3): 343–73. https://doi.org/10.1080/15387216.2016.1243065

Jargalsaikhan, D. 2015. 'Bidnii setgelgeenii khotsrogdol ba ediin zasgiin erkh chölöö' [Backwardness of our understanding and the freedom of economy]. Paper presented at the For Mining without Populism forum. Ulaanbaatar, 3 September.

Jargalsaikhan, Mendee. 2018. 'Mongolia's Dilemma: A Politically Linked, Economically Isolated Small Power'. In *International Relations in Asia's Northern Tier: Sino-Russian Relations, North Korea, and Mongolia*, edited by G. Rozman and S. Radchenko, 157–75. Singapore: Palgrave Macmillan.

Jigmeddorj, E. 2015. 'Mongolchuudyn baigali khamgaalakh ulamjlal: Altny kharuulyn jisheegeer' [Mongolians tradition of protecting nature: On the example of gold patrol]. Paper presented at the 36th International conference of the Korean Association for Mongolian Studies. Seoul, March 2015.

Joffé, George, Stevens, Paul, George, Tony, Lux, Jonothan and Searle, Carol. 2009. 'Expropriation of Oil and Gas Investments: Historical, Legal and Economic Perspectives in a New Age of Resource Nationalism', *Journal of World Energy Law & Business* 2(1): 3–23. https://doi.org/10.1093/jwelb/jwn022

Kaplonski, Christopher. 2005. 'The Case of the Disappearing Chinggis Khaan: Dismembering the Remembering', *Ab Imperio* 4: 147–73.

Khan, Almaz. 1996. 'Who Are the Mongols? State, Ethnicity and the Politics of Representation in the PRC'. In *Negotiating Ethnicities in China and Taiwan*, edited by Melissa J. Brown, 125–59. Berkeley: Institute of East Asian Studies, University of California.

Khazanov, Anatoly M. 1984. *Nomads and the Outside World*. Trans. Julia Crookenden. Cambridge: Cambridge University Press.

Kirkpatrick, Noel. 2018. 'What is cloud seeding, and does it really work?' Last accessed 10 March 2019. www.mnn.com/earth-matters/climate-weather/stories/what-cloud-seeding

Kirsch, Stuart. 2014. *Mining Capitalism: The Relationship between Corporations and Their Critics*. Berkeley: University of California Press.

Koch, Natalie and Perreault, Tom. 2018. 'Resource Nationalism', *Progress in Human Geography* 43(4): 611–31. https://doi.org/10.1177/0309132518781497

Kolås, Åshild. 2014. 'Degradation Discourse and Green Governmentality in the Xilinguole Grasslands of Inner Mongolia', *Development and Change* 45(2): 308–28. https://doi.org/10.1111/dech.12077

Konagaya, Yuki and Lkhagvasuren Ichinkhorloo. 2014. 'Choijingiin Khurts, Geologi uul uurkhain yamny said asan: 2013 ony 3 sard yariltsasan' [Choijingiin Khurts, Geologist, Formerly the Minister of Mining: Interview conducted in March 2013]. *Senri Ethnological Reports: Mongolia's Transition from Socialism to Capitalism: Four Views* 121(0): 93–180. www.minpaku.ac.jp/english/research/activity/publication/other/ser/121

Kradin, Nikolay. 2012. *Khünnü Ezent Uls* (Xiongnu Empire). N. Ganbat trans. Ulaanbaatar: Soyombo Printing.

Krastev, Ivan. 2004. *Shifting Obsessions: Three Essays on the Politics of Anticorruption*. Budapest and New York: Central European University Press.

Kretzschmar, Gavin L., Kirchner, Axel. and Sharifzyanova, Liliya. 2010. 'Resource Nationalism: Limits to Foreign Direct Investment', *Energy Journal* 31(2): 27–52. https://ssrn.com/abstract=1317395 or http://dx.doi.org/10.2139/ssrn.1317395

Larson, Christina. 2014. 'Prominent Mongolian Environmentalist Given 21-Year Jail Sentence for "Terrorism"'. Last accessed 29 January. www.bloomberg.com/news/articles/2014-01-29/mongolian-environmentalists-get-21-years-in-prison-for-terrorism

Leiss, William. 1994. *The Domination of Nature*. Montreal: McGill-Queen's University Press.

Lhamsuren, Munkh-Erdene. 2006. 'The Mongolian Nationality Lexicon: From the Chinggisid Lineage to Mongolian Nationality (From the Seventeenth to the Early Twentieth Century)', *Inner Asia* 8(1): 51–98.

Li, Fabiana. 2015. *Unearthing Conflict: Corporate Mining, Activism, and Expertise in Peru*. Durham, NC and London: Duke University Press.

Linebaugh, Peter. 2008. *The Magna Carta Manifesto: Liberties and Commons for All*. Berkeley: University of California Press.

Magvanjav, Bazaryn and Tsogtbaatar, Choijinjavyn. 2017. *Mongol Ulsyn Uul Uurkhai 95: Mongolyn uul uurkhain tüükhen khögjliin jim* (Mining of Mongolia 95: Routes of the historical development of mining in Mongolia). Ulaanbaatar: Inter Press.

Mair, Jonathan. 2012. 'GDAT 2012: Debating "neoliberalism"'. Last accessed 5 December 2019. http://jonathanmair.com/gdat-2012-the-concept-of-neoliberalism-has-become-an-obstacle-to-the-anthropological-understanding-of-the-twenty-first-century/

Maniruzzaman, A.F.M. 2009. 'The Issue of Resource Nationalism: Risk Engineering and Dispute Management in the Oil and Gas Industry', *Texas Journal of Oil, Gas, and Energy Law* 5(1): 79–108. https://ssrn.com/abstract=1985171

Marin, Andrei. 2010. 'Riders Under Storms: Contributions of Nomadic Herders' Observations to Analysing Climate Change in Mongolia', *Global Environmental Change* 20(1):162–76. https://doi.org/10.1016/j.gloenvcha.2009.10.004

Marsh, Peter K. 2009. *The Horse-Head Fiddle and the Cosmopolitan Reimagination of Tradition in Mongolia*. London: Routledge.

Marzluf, Phillip. 2015. 'The Pastoral Home School: Rural, Vernacular, and Grassroots Literacies in Early Soviet Mongolia', *Central Asian Survey* 34(2): 204–18. https://doi.org/10.1080/02634937.2014.991611

Mineral Resources and Petroleum Authority of Mongolia. 2015. '*Alt 2025' hötölböriin suuri sudalgaany tailan* ['Gold 2025' programme baseline research report], accessed 27 March 2018. https://mrpam.gov.mn/public/pages/66/Алт%202025%20хөтөлбөрийн%20суурь%20судалгааны%20тайлан.pdf

Mitchell, Timothy. 2002. *Rule of Experts: Egypt, Techno-Politics, Modernity*. Berkeley: University of California Press.

Mongol News. 2011a. 'Ekh oron nam khaachiv?' [Where did the Homeland Political Party go?] Last accessed 11 April. http://mongolnews.mn/dwq.

Mongol News. 2011b. '"Urt nertei" khuuliig shine tösöl avrakh uu alakh uu' [Will the new project save or kill the 'law with the long name']. Last accessed 15 February 2019. http://mongolnews.mn/cic.

Mongolian Mining Journal. 2008. 'Mongol ulsyn tösviin zarlaga Genetiin ashgiin tatvaryn khuuli batlagdsan üyes ekhlen ogtsom buyu jild dundjaar 56 khuviar nemegdjee' [Since the approval of the windfall tax law GDP of Mongolia dramatically increased average 56% per year]. Last accessed 11 August 2019. www.mongolianminingjournal.com/content/15310.shtml.

Mongyol ulus-un yeke qural (State Great Khural [Parliament] of Mongolia). 1924. Bügüde nayiramdaqu monggol arad ulus-un ündüsün qauli [Constitution of the People's Republic of Mongolia]. Assent 29 November.

Mongol Ulsyn Ikh Khural (State Great Khural [Parliament] of Mongolia). 1994. 'Ashigt Maltmalyn Tukhai' [Minerals Law]. Assent 5 June.

Mongol Ulsyn Ikh Khural (State Great Khural [Parliament] of Mongolia). 1995. 'Baigali orchnyg khamgaalakh tukhai khuuli' [Environmental Protection Law]. Assent 30 March. www.legalinfo.mn/law/details/8935.

Mongol Ulsyn Ikh Khural (State Great Khural [Parliament] of Mongolia). 1997. 'Ashigt Maltmalyn Tukhai' [Minerals Law]. Assent 5 July. http://legalinfo.mn/law/details/7069

Mongol Ulsyn Ikh Khural (State Great Khural [Parliament] of Mongolia). 2002. 'Erüügiin Khuuli' [Criminal Law]. Assent 3 January. www.legalinfo.mn/law/details/12172

Mongol Ulsyn Ikh Khural (State Great Khural [Parliament] of Mongolia). 2006. 'Ashigt Maltmalyn Tukhai (Shinechilsen nairuulga)' [Minerals Law (Amended)]. Assent 8 July. www.legalinfo.mn/law/details/63?lawid=63

Mongol Ulsyn Ikh Khural (State Great Khural [Parliament] of Mongolia). 2009. 'Gol, mörnii ursats büreldekh, usny san bükhii gazryn khamgaalaltyn büs, oin san bükhii gazart ashigt maltmal khaikh, ashiglakhyg khoriglokh tukhai' [Law to prohibit mineral exploration and mining operations at headwaters of rivers, protected zones of the water reservoir and forest area]. Assent 16 July. www.legalinfo.mn/law/details/223

Mongol Ulsyn Ikh Khural (State Great Khural [Parliament] of Mongolia). 2010. *Ündesnii ayulgüi baidlyn üzel barimtlal* [National security concept]. Assent 15 July. www.openforum.mn/res_mat/res_mat-65.pdf

Moreno, Kasia. 2014. 'Regulatory Environment Has More Impact on Business Than the Economy, Say U.S. CEOs'. Last accessed 12 August 2019. www.forbes.com/sites/forbesinsights/2014/08/12/regulatory-environment-has-more-impact-on-business-than-the-economy-say-u-s-ceos/-7fa83bb4684d

Munkherdene, Gantulga. 2018. 'The Formation and Distribution of Procapitalist Perspectives in Mongolia', *Central Asian Survey* 37(3): 372–85. https://doi.org/10.1080/02634937.2018.1499609

Munkh-Erdene, Lhamsuren. 2012. 'Mongolia's Post-Socialist Transition: A Great Neoliberal Transformation'. In *Mongolians after Socialism: Politics, Economy, Religion*, edited by Bruce M. Knauft, Richard Taupier, and Lkham Purevjav, 61–7. Ulaanbaatar: Admon Press.

Murray-Li, Tanya. 2007. 'Practices of Assemblage and Community Forest Management', *Economy and Society* 36(2): 263–93. https://doi.org/10.1080/03085140701254308

Myadar, Orhon. 2011. 'Imaginary Nomads: Deconstructing the Representation of Mongolia as a Land of Nomads', *Inner Asia* 13(2): 335–62. https://doi.org/10.1163/000000011799297654

Myadar, Orhon and Jackson, Sara. 2018. 'Contradictions of Populism and Resource Extraction: Examining the Intersection of Resource Nationalism and Accumulation by Dispossession in Mongolia', *Annals of the Association of American Geographers*. https://doi.org/10.1080/24694452.2018.1500233

Nasanbaljir, J. 1964. *Ar mongoloos manj chin ulsad zalguulj baisan alba 1691–1911* [Tributes of Mongolia to the Manchu Qing 1691–1911]. Ulaanbaatar: Ulsyn khevleliin gazar.

Navaro-Yashin, Yael. 2002. *Faces of the State: Secularism and Public Life in Turkey*. Princeton, NJ and Oxford: Princeton University Press.

New Zealand Parliament. 1994. 'Resource Management Amendment Act 1994'. Assent November 14. www.legislation.govt.nz/act/public/1994/0105/latest/DLM341408.html

Ninjsemjid. L. 2012. 'Litsenziin khorig tavigdaj nair bolno gev üü, yerönkhiilögch öö?' [President, hear that there will be a celebration upon release of the licence ban?]. *Niigmiin Toil*. Last accessed 21 December 2019. www.sonin.mn/news/politics-economy/11096.

Odonchimeg, L. 2015. 'Ch. Saikhanbileg: Mongolyn talyn khuviig 54.9 bolgokh bükh tokhirol-tsoog khiisen' [Ch. Saikhanbileg: Have made of the negotiations to make the Mongolian percentage of shares 54.9]. *Ikon News*, 21 May 2015. https://ikon.mn/n/h28

Ödriin Sonin [Daily News]. 2007. 'Altny uukhainuud baigali orchny nökhön sergeelt khiideggüi' [Gold mining companies fail to engage in environmental rehabilitation]. 30 August.

Ong, Aihwa. 2006. *Neoliberalism as Exception: Mutations in Citizenship and Sovereignty*. Durham, NC: Duke University Press.

Otgonsuren, S. 2011. 'Margaash "baryn bazalt khiine" gev' [Said 'there will be a tiger squeezing' tomorrow]. Last accessed 3 June 2019. http://old.news.mn/r/69457.

Papadopoulos, Dimitris. 2010. 'Insurgent Posthumanism', *Ephemera: Theory and Politics in Organization* 10(2):134–51.

Peet, Richard. 2003. *Unholy Trinity: The IMF, World Bank and WTO*. New York: Zed Books.

Plueckhahn, Rebekah and Bumochir, Dulam. 2018. 'Capitalism in Mongolia: Ideology, Practice and Ambiguity', *Central Asian Survey* 37(3): 341–56. https://doi.org/10.1080/02634937.2018.1510600

Polanyi, Karl. 2001. *The Great Transformation: The Political and Economic Origins of Our Time*. Boston: Beacon Press.

Qian, Sima. 1993. *Records of the Grand Historian of China*. Burton Watson trans. New York: Colombia University Press.

Rajak, Dinah. 2011. *In Good Company. An Anatomy of Corporate Social Responsibility*. Palo Alto, CA: Stanford University Press.

Raman, Ravi and Lipschutz, Ronnie eds. 2010. *Corporate Social Responsibility: Comparative Critiques*. London: Palgrave Macmillan.

Reeves, Jeffrey. 2011. 'Resources, Sovereignty, and Governance: Can Mongolia Avoid the "Resource Curse"?' *Asian Journal of Political Science* 19(2): 170–85. https://doi.org/10.1080/02185377.2011.600165

Richardson, Tanya and Weszkalnys, Gisa. 2014. 'Introduction: Resource Materialities', *Anthropological Quarterly* 87(1): 5–30. doi:10.1353/anq.2014.0007

Riley, Charles. 2012. 'World's Best Economies'. Last accessed 13 August. https://money.cnn.com/gallery/news/economy/2012/08/13/worlds-best-economies/5.html

Ross, Michael. 1999. 'Political Economy of the Resource Curse', *World Politics* 51(2): 297–322. https://doi.org/10.1017/S0043887100008200

Rossabi, Morris. 2005. *Modern Mongolia: From Khans to Commissars to Capitalists*. Berkeley: University of California Press.

Rosser, Andrew. 2006. 'IDS Working Paper 268: The political economy of the Resource Curse: A Literature Survey'. Last accessed 1 January 2019. www.ids.ac.uk/files/WP268.pdf

Sachs, Jeffrey D. and Warner, Andrew M. 2001. 'Natural Resources and Economic Development: The Curse of Natural Resources', *European Economic Review* 45(4–6): 827–38. http://dx.doi.org/10.1016/S0014-2921(01)00125-8

Sahlins, Marshall. 1972. *Stone Age Economics*. Chicago: Aldine-Atherton.

Sanchata, Mariko. 2012. 'Mongolia aims to woo foreign investors'. Last accessed 31 October 2019. http://online.wsj.com/article/SB10001424052970204712904578090282738517610.html

San Juan, Epifanio. 2002. 'Nation-State, Postcolonial Theory, and Global Violence', *Social Analysis* 46(2): 11–32.

Sanchir, Jargalsaikhan. 2016. 'Bayalgiin ündesnii khyanalt' [National control on resources], *Shinjeech* 1: 54–79.

Sanders, Alan. 1996. 'Foreign Relations and Foreign Policy'. In *Mongolia in Transition: Old Patterns and New Challenges*, edited by Ole Bruun and Ole Odgaard, 217–52. Richmond: Curzon.

Sansar, B. 2015. '"Altan Dornod Mongol" kompani khamgiin ikh ashiglaltyn litsenz ezemshdeg' ['Altan Dornod Mongol' company possesses the largest number of extraction licences]. Last accessed 10 November. www.olloo.mn/n/22511.html

Sassen, Saskia. 1996. *Losing Control? Sovereignty in an Age of Globalization*. New York: Columbia University Press.

Sawyer, Suzana. 2004. *Crude Chronicles: Indigenous Politics, Multinational Oil and Neoliberalism in Ecuador*. Durham, NC: Duke University Press.

Semuun, B. 2013. 'Bayalag ünegüidsen shaltgaan' [The reason for resources to become valueless]. Last accessed 18 September 2019. www.news.mn/?id=143134

Shaiko, Ronald G. 1987. 'Religion, Politics and Environmental Concern: A Powerful Mix of Passions', *Social Science Quarterly* 68(2): 244–62.

Shever, Elana. 2012. *Resources for Reform: Oil and Neoliberalism in Argentina*. Stanford, CA: Stanford University Press.

Shiirev, Ts. 2017. 'Noyon uulyg yagaad ukhaj bolokhgüi talaarkh olny tanil khümüüsiin setgegdlees' [From the impression of some famous people about why Noyon mountain cannot be extracted]. Last accessed 8 February 2019. http://unuudur.mn/article/94583

Shimamura, Ippei. 2014. 'Ancestral Spirits Love Mining Sites: Shamanic Activities Around a Copper-Gold Mining Site in Mongolia', *Inner Asia* 16(2): 393–408. https://doi.org/10.1163/22105018-12340025

Shore, Cris and Wright, Susan. 1997. *The Anthropology of Politics. Perspectives on Governance and Power*. London: Routledge.

Shore, Cris, Wright, Susan and Pero, Davide, eds. 2011. *Policy Worlds: Anthropology and the Analysis of the Contemporary Power*. New York and Oxford: Berghahn.

Simonov, Eugine. 2013. 'Short History of the Law with the Long Name'. Last accessed 9 November 2019. www.transrivers.org/documents/rivers-and-mining/the-short-history-of-the-law-with-long-name/

Smith, Jessica R. 2013. 'The Politics of Pits and the Materiality of Mine Labor: Making Natural Resources in the American West', *American Anthropologist* 115(4): 582–94. doi:10.1111/aman.12050

Sneath, David. 2001. 'Notions of Rights over Land and the History of Mongolian Pastoralism', *Inner Asia* 3(1): 41–58. https://doi.org/10.1163/146481701793647750

Sneath, David. 2002. 'Mongolia in the "Age of the Market": Pastoral Land-use and the Development Discourse'. In *Markets and Moralities: Ethnographies of Postsocialism*, edited by C. Humphrey and R. Mandel, 101–210. Oxford: Berg.

Sneath, David. 2003. 'Land Use, the Environment and Development in Post-Socialist Mongolia', *Oxford Development Studies* 31(4): 441–59. https://doi.org/10.1080/1360081032000146627

Sneath, David. 2004. 'Property Regimes and Sociotechnical Systems: Rights Over Land in Mongolia's "Age of the Market"'. In *Property in Question: Value Transformation in the Global Economy*, edited by Katherine Verdery and Caroline Humphrey, 161–84. Oxford and New York: Berg.

Sneath, David. ed. 2006. *Imperial Statecraft: Political Forms and Techniques of Governance in Inner Asia, Sixth-Twentieth Centuries*. Bellingham: WA Centre for East Asian Studies, Western Washington University Press.

Sneath, David. 2007. *The Headless State: Aristocratic Orders, Kinship Society and Misinterpretations of Nomadic Inner Asia*. New York: Columbia University Press.

Sneath, David. 2010. 'Political Mobilization and the Construction of Collective Identity in Mongolia', *Central Asia Survey* 29(3): 251–67. https://doi.org/10.1080/02634937.2010.518009

Sneath, David. 2012. 'The "Age of the Market" and the Regime of Debt: The Role of Credit in the Transformation of Pastoral Mongolia', *Social Anthropology/Anthropologie Sociale* 20(4): 458–73. https://doi.org/10.1111/j.1469-8676.2012.00223.x

Snow, Keith Harmon. 2010. 'The naked face of capitalism: Goldman Prizewinner Shoots up Foreign Mining Firms in Mongolia: Western deception and the extinction of the nomads'. Last accessed February 2011. https://survivingcapitalism.blogspot.com/2010/12/naked-face-of-capitalism-goldman.html

Snow, Keith Harmon. 2014. 'The naked face of capitalism: Goldman Prizewinner Gets 21 Years for Resistance to Genocide: Foreign Mining, State Corruption and Genocide in Mongolia'. Last accessed 9 February 2019. www.consciousbeingalliance.com/2014/02/goldman-prizewinner-gets-21-years-for-resistance-to-genocide/

Sodbaatar, Ya. 2013. *Mongolyn uul uurkhain tüükh* [History of mining in Mongolia]. Ulaanbaatar: Admon Press.

Song, Ligang and Woo, Wing Thye eds. 2008. *China's Dilemma: Economic Growth, The Environment and Climate Change*. Canberra: Asia Pacific Press, Brookings Institution Press and Social Science Academic Press (China).

Standartchillyn ündesnii zövlol (National Council of Standardisation). 2012. *Mongol kiril üsgiin latin galig* [Latin transliteration of the Mongolian Cyrillic]. Approved 16 February. www.estandard.gov.mn/index.php?module=standart&cmd=standart_desc&sid=2579

Stanway, David and Edwards, Terrence. 2012. 'Resource-rich Mongolia plays populist card in run-up to polls'. Last accessed 11 August 2019. www.moneycontrol.com/news/business/wire-news/-1824877.html

Steger, Manfred B. and Roy, Ravi K. 2010. *Neoliberalism: A Very Short Introduction*. Oxford: Oxford University Press.

Stépanoff, Charles, Marchina, Charlotte, Fossier, Camille and Bureau, Nicolas. 2017. 'Animal Autonomy and Intermittent Coexistences: North Asian Modes of Herding', *Current Anthropology* 58(1): 57–81. www.journals.uchicago.edu/doi/full/10.1086/690120?mobileUi=0

Sternberg, Troy. 2010. 'Unraveling Mongolia's Extreme Winter Disaster of 2010', *Nomadic Peoples* 14(1): 72–86. https://doi.org/10.3167/np.2010.140105

Stiglitz, Joseph E., Sen, Amartya, and Fitoussi, Jean-Paul. 2010. *Mismeasuring Our Lives: Why GDP Doesn't Add Up*. New York: New Press.

Swire, Mary. 2009. 'Mongolia to rescind mining windfall tax'. Last accessed 28 August 2019. www.tax-news.com/news/Mongolia_To_Rescind_Mining_Windfall_Tax____38740.html

Tanaka, Katsuhiko. 2002. 'Kokka naku shite minzoku ha ikinokoreru ka: Buriyaato Mongoru no chishikijintachi' [Can a Nation Survive without a State? Intellectuals of Buryat Mongols], in E. Kuroda (ed.), Minzoku no undou to shidousha tachi [National Movements and Their Leaders]: 74–95. Tokyo: Yamakawa Shuppan.

Taussig, Michael. 1992. *The Nervous System*. New York and London: Routledge.

Terenguto, A. 2004. 'Relationship Between Man and Nature: A Hermeneutical Approach to Interpreting the Affective Thinking of the Mongolian People', *Inner Asia* 6(1): 81–93. https://doi.org/10.1163/146481704793647243

Tolson, Michael. 2014. 'Eco-warrior or eco-terrorist? Mongolia jails environmentalist for 21 years'. Last accessed 28 January. https://asiancorrespondent.com/2014/01/mongolia-tsetsegee-munkhbayar-jail/-Ew7EPjeEvk6t44Av.97

Tömör Temür Tamerlan. 2018. '1990 ony khuvisgal khuvisgal bish baisan' [The 1990 revolution was not a revolution]. Facebook, 28 May 2018. www.facebook.com/groups/1377355269164330/permalink/2163713927195123/

Torbati, Yeganeh. 2016. 'Kerry Hails Mongolia as "Oasis of Democracy" in Tough Neighbourhood'. www.reuters.com/article/us-usa-mongolia-idUSKCN0YR02T.

Trading Economics. 2018. 'Mongolia GDP Growth Rate YoY'. Last accessed 12 September 2018. https://tradingeconomics.com/mongolia/gdp-growth.

Tse, Pui-Kwan. 2007. *2006 Minerals Year Book: Mongolia*. Last accessed 18 September 2018. https://s3-us-west-2.amazonaws.com/prd-wret/assets/palladium/production/mineral-pubs/country/2006/myb3-2006-mg.pdf

Tsenddoo, B. 2015. *Irgenshliin zamd: Mongolyn soyolyn almanakh (On the way towards civilization: Almanac of Mongolian culture)*. Ulaanbaatar: Nepko Publishing.

Tseren, P. 1996. 'Traditional Pastoral Practice of the Oirat Mongols and Their Relationship with the Environment'. In *Culture and Environment in Inner Asia. Volume 2: Society and Culture*, edited by C. Humphrey and D. Sneath, 147–59. Cambridge: White Horse Press.

Tsetsentsolmon, B. 2014. 'The "Gong Beat" against the "Uncultured": Contested Notions of Culture and Civilization in Mongolia', *Asian Ethnicity* 15(4): 422–38. https://doi.org/10.1080/14631369.2014.947060

Tsing, Anna. 2005. *Friction: An Ethnography of Global Connection*. Princeton, NJ: Princeton University Press.

Tsing, Anna. 2012. 'Unruly Edges: Mushrooms as Companion Species', *Environmental Humanities* 1: 141–54. www.environmentandsociety.org/node/5415

Tsogzolmaa, J. 2010. 'Geologi uul uurkhain üildveriin said asan, doctor Ch. Khurts: Khuulia khuivaldaand tokhiruulsanaas bid Luugiin golyn tolgoin ordoo aldsan shüü dee' [Ch. Khurts, doctor and former minister of geology and mining productions: We lost our deposit in Luugiin Tolgoi for fitting our law to conspiracy]. Last accessed 11 August 2019. http://news.gogo.mn/r/67550.

Tuya, S. and Battomor, B. 2012. *Uul uurkhai mongol orond (Mining in Mongolia)*. Ulaanbaatar: Jicom Press.

Ul-Oldokh, Ch. 2014. 'S. Narangerel: Ts. Munkhbayar nar ingej shiitgüülekh uchirgüi' (S. Narangerel: Ts. Munkhbayar and others are not supposed to be sentenced like this). Last accessed 24 January 2020. www.sonin.mn/news/politics-economy/23516

Upton, Caroline. 2010. 'Nomadism, Identity and the Politics of Conservation', *Central Asian Survey* 29(3): 305–19. https://doi.org/10.1080/02634937.2010.518010

Upton, Caroline. 2012. 'Mining, Resistance and Pastoral Livelihoods in Contemporary Mongolia'. In *Change in Democratic Mongolia: Social Relations, Health, Mobile Pastoralism, and Mining*, edited by Julian Dierkes, 223–49. Leiden: Brill.

Vanoli, André. 2005. *A History of National Income Accounting*. Amsterdam: IOS Press.

Venzon, Cliff. 2018. 'Flare-up of resource nationalism burns miners across Asia'. Last accessed 11 August 2019. https://asia.nikkei.com/Business/Markets/Commodities/Flare-up-of-resource-nationalism-burns-miners-across-Asia

Verdery, Katherine and Humphrey, Caroline. 2004. 'Introduction: Raising Questions about Property'. In *Property in Question: Value Transformation in the Global Economy*, edited by Katherine Verdery and Caroline Humphrey, 1–29. Oxford and New York: Berg.

Vitebsky, Piers. 1995. 'From Cosmology to Environmentalism: Shamanism as Local Knowledge in a Global Setting'. In *Counterworks: Managing the Diversity of Knowledge*, edited by R. Fardon, 182–203. London: Routledge.

Vivoda, Vlado. 2009. 'Resource Nationalism, Bargaining and International Oil Companies: Challenges and Change in the New Millennium', *New Political Economy* 14(4): 517–34. https://doi.org/10.1080/13563460903287322

Vladimirtsov, B. I. 1971. *Comparative Grammar of the Mongolian Written Language and the Khalkha Dialect*. Cuckfield: Gregg Publishing

Walker, Danny. 2001. 'Placer Gold Mining in Mongolia – the New Zealand Way', *World Placer Journal* 2(1): 36–49.

Watson, Nicholas. 2012. 'Mongolia: Mine ownership gets political'. Last accessed 11 August 2019. www.ft.com/content/91195805-14ed-3839-8b8c-a72ea6609cd4

Watts, Michael. 2003. 'Development and Governmentality', *Singapore Journal of Tropical Geography* 24(1): 6–34. https://doi.org/10.1111/1467-9493.00140

Watts, Michael. 2004. 'Resource Curse? Governmentality, Oil and Power in the Niger Delta, Nigeria', *Geopolitics* 9(1): 50–80. https://doi.org/10.1080/14650040412331307832

Welker, Marina. 2014. *Enacting the Corporation: An American Mining Firm in Post-Authoritarian Indonesia*. Berkeley: University of California Press.

Weszkalnys, Gisa. 2011. 'Cursed Resources, or Articulations of Economic Theory in the Gulf of Guinea', *Economy and Society* 40(3): 345–72. https://doi.org/10.1080/03085147.2011.580177

Williams, Dee Mack. 2000. 'Representation of Nature on the Mongolian Steppe: An Investigation of Scientific Knowledge Construction', *American Anthropologist* 102(3): 503–19. https://doi.org/10.1525/aa.2000.102.3.503

Wilson, Jeffrey. 2015. 'Understanding Resource Nationalism: Economic Dynamics and Political Institutions', *Contemporary Politics* 21(4): 399–416. https://doi.org/10.1080/13569775.2015.1013293

World Bank. 2004. 'Mongolia: Mining Sector Sources of Growth Study'. Last accessed 15 September 2018. https://openknowledge.worldbank.org/handle/10986/14397

World Bank. 2012. 'What is behind Mongolia's Economic Boom?' Last accessed 11 August 2019. www.worldbank.org/en/news/video/2012/02/28/what-behind-mongolia-economic-boom

Yeh, Emily. 2005. 'Green Governmentality and Pastoralism in Western China: "Converting Pastures to Grasslands"', *Nomadic Peoples* 9(1): 9–30. https://doi.org/10.3167/082279405781826164

Zulbayar, B. 2015. 'Today's dilemma: Short term tactics or long term strategy?' Last accessed 11 August 2019. http://ubpost.mongolnews.mn/?p=16531

Index

accessing the state 112
activism 82, 88, 140
advocacy organisations 81–3, 90, 95, 99
AGR Limited 72
Albright, Madeleine 28–9
Algaa Namgar 32–7, 43–4, 54, 56, 67
Ali, Saleem 2, 40–2, 49
Altangerel, P. 110
Amarjargal Rinchinnyam xviii
Amarsanaa, S. 88, 91
Anderson, Desaix 28
anthropogenic forces 131–2
Appel, Hannah 23–4, 37, 48, 139
The Asia Foundation (TAF) 30, 89, 94–9, 121
Asian Development Bank (ADB) 26–7, 30
AUM (gold mining company) 125
Avirmed, S. 43

Badamsambuu, G. 103
Baker, James 28
balancing action taken by the state 53–9
Baljinnyam, B. 116
Ballard, Chris 62–3
Banks, Glenn 62–3
Bargh, Maria 139
Barrington, Lowell 9–10, 12, 22, 102, 117
Batbaatar (company) 77–9
Batbayar Nyamjav 56
Batbayar Jargalan 56
Batbold Sukhbaatar 108–9
Bat-Erdene Badmaanyambuu 105
Battulga Khaltmaa 43

Battuvshin, B. 88
bayalag büteegchid 17
bayalgiin ündesnii khyanalt 41; *see also* national resource control
Bayarsaikhan Namsrai 64, 85, 88–91, 95–9, 103
Bear, Laura 61
Bebbington, Anthony 3–4, 11, 62, 90, 94, 102
berry trees, planting of 98–9
Blaser, Mario 3
Boas, Taylor 5–6
Bogd Khan 51
Bolor, D. 25
Boroo Gold 34–6
Bremmer, Ian 40
Buddhism 47
Byambaa, J. 43
Byambabaabar Ichinkhorloo 125–6
Byambajav, Dalaibuyan 33, 81, 99
Byambasuren Dash xvi, 16, 26–31, 36

de la Cadena, Marisol 3
'camels' protest (2002) 121
Canadian Centerra Gold Inc. xii–xiii
Canadian Mining and Energy Corporation (Cameco) 72
Centerra Gold (company) 72, 108–9
Chandmani Dambabazar 87–8
Childs, John 2, 41–2
China 8–9, 21, 24–7, 31–2, 137–8
China Shenhua Energy 62
Chinbat Lhagva ix, 63, 67–72, 130
Chinggis Khan 101, 107, 116, 118–19, 129
Choijilsuren Battogtokh 137
civic principles 9–10

civil society 88, 96–9, 140
climate change 123
Clinton, Hillary 28–9
Cold Gold Mongolia (CGM) 64, 71–9, 82
Collier, Paul xv
'commoning' process 3
community-based organisations 81–2
compensation payments 76, 106, 110
constitutions of Mongolia 12, 52, 54, 109, 116–17
Cordero, Raúl R. xiv
Coronil, Fernando 101
Croft, Layton 89–90, 94–7
crystal extraction 46

Dagvadori, D. 132
Daly, Herman E. xiv
Dashdemberel Ganbold 30, 104–5, 109, 115, 125, 136
Da Silva, Luis xiv
Davaa, N. 107
Davaanyam, P. 107
'debt trap' 137–8
deification and de-deification of the state 101–2, 110
desertification 123
Dierkes, Julian 2, 41–2
discursive resources 58–9, 90
donor organisations 17–18, 89, 95–9
Dulam, S. 132

Ekins, Paul xiv
election rigging 108
elite groups 22, 87
Elizabeth II, Queen 31
Empson, Rebecca xv, xviii, 2, 32, 41–3, 53
Enkhbold, Z. 91
Enkhjargal Dandinbazar 34
Enkhsaikhan Mendsaikhan 33, 65
Enkhsaikhan Oomoo 55
environmental protection xi, xiv–xv, 9–12, 18–19, 53, 68–9, 74, 84–5, 91, 98, 103, 117–19, 124, 130–3, 136, 139
Environmental Protection Law 52, 105–10
Equatorial Guinea 23, 48
Erdenebaatar, I. 65
Erdenebat, B. 65
Erdenet (company) 56, 106
Erel (company) 63–7, 70–1, 76, 86–7, 91, 93, 103
Eriksen, Thomas Hylland 5
Erkhis Mining 73–4
ethnicity 9–12
ethnography 1, 101–2

Fire Nation (FN) coalition 107–15, 128
food production 70
Fortuna's daughter, law of 56
Fratkin, Elliot 123
free market economy 46, 58
Friedman, Milton 45

Galdan Khan 49
Ganbold, Misheelt 2, 40–2, 49
Gans-Morse, Jordan 5–6
Gantulga, N. 125–6
Gatsuurt (company) xii, 63–4, 67–71, 73, 76
Gilberthorpe, Emma 13, 63
Gluchowski, Peter 88
gold mining 38, 56–7, 60–4, 68–72, 75–9, 85, 93
gold patrol 50
Gold Programmes 32, 58
Goldin, Ian xiv
Goldman Prize 84, 93–5, 111, 114, 126–9
Gorbachev, Mikhail 28
Graeber, David 29
greed 47
Group for Debates in Anthropological Theory (GDAT) 6

Habermas, Jürgen 102
Hannam, Ian 123
Haraway, Donna J. 133
Harvey, David 29–30
Hatcher, Pascale xvi
Henke, Michael 88–9
High, Mette xv, 52, 65, 95
homeland claims 22
'horse' operation (2011) 110–11
Humphrey, Caroline xv, 26, 102, 124
hunter-gatherer societies 122
Hürelbaatar, A. 102

imagined communities 4, 36, 101
imprisonment 114, 117–19, 125, 127
independence of Mongolia 21–6
indigenisation 4–5, 7–8
'indigenous', two meanings of 5
International Monetary Fund (IMF) 26–30, 57, 137
internationalisation 93–4
Ivanhoe Mines 57, 95, 97, 108

Jackson, Sara 22, 41
Jacques, Jean-Sebastien 40
Jargalsaikhan Dugar 11–12, 33, 35, 54
Javkhlan Samand 107
Johnston, Robert 40

Kerry, John 29
Khan Shijir (company) 73–4

156   INDEX

Khashchuluun Chuluundorj 34
Khüitnii Am mine 74–7
Khurts Choijin 16, 43–7, 50–5
Kirsch, Stuart xv, 24, 36, 61, 63
Koch, Natalie 2, 41–2
Konrad-Adenauer Stiftung (KAS)
  88–9, 99
Kyrgyzstan 40

Laos xvi
law of nature 132
Law of Petition 118
'law with the long name' 67, 69–70,
  82, 104, 119, 125, 136; *see also*
  prohibition by law of 'mineral
  exploration and mining operations at
  headwaters of rivers, protected zones
  of the water reservoir and forest area'
'legitimate state' concept 118–19
Lhamsuren, Munkh-Erdene 7, 9, 25,
  30, 118
Libya 23
licences for mining 34–5, 54–5, 66–7,
  70–1, 79 106; international 40
'lifeworld' concept 102
Linebaugh, Peter 3
Li Peng 27
Lkhagvadorj, O. 107
lobbying 91, 119
local communities 13–14
Lundeejantsan, D. 91

Mair, Jonathan 5–6
malchin identity 125, 128, 130, 133
marketisation 7–8
Mearns, Robin 123
Mineral Resources and Petroleum
  Authority of Mongolia
  (MRPAM) 34
Minerals Laws 33–6, 44, 47, 54–6
mining industry 13–15, 51–2, 60–2,
  77–9, 108–9, 138; *see also* gold
  mining
mining technology 68
minority groups 22
Mitchell, Timothy 30
mobile phones, use of 105
mobilisations 82–4
Modu Chanyu 49
Mongol Empire 118
Mongolian Democratic Party
  (MDP) 84
Mongolian Mining Journal 57
Mongolian National Mining Association
  (MNMA) 11
Mongolian Nature Protection Coalition
  (MNPC) 96, 103, 109, 121
Mongolian People's Party (MPP) 84
mongolisation 5, 7–8

Mongolor project 50–1
Mönkh Tenger 118–19, 130–2
Motherland Party 65–7, 91–2
multinational corporations 61–2;
  image of 62
multipartite relationships 1, 13
Munkhbayar Tsetsegee xi, 12, 18,
  65–6, 76, 84–7, 90–9, 102–19,
  121–33
Munkherdene, Gantulga 26
My Mongol Homeland coalition 92
Myadar, Orhon 41

nation-building 7–9, 17, 23–4, 37, 60,
  139–41
nation-states 6, 12
national identity 122–3, 133
national resource control 41–3, 48–56
nationalisation 91–4
nationalism xviii, 2–3, 8–10, 15–18,
  22–3, 37, 49, 58, 60–5, 70, 102, 117,
  136–9; civic 10; 'good' and 'bad'
  10, 46–7, 60; *see also* 'resource
  nationalism'
'navigation' (Bear) 61, 64–5, 71, 74–8
neoliberalisation 4–6, 8, 140
neoliberalism xviii, 3–10, 22, 24, 41,
  45, 137–40
New Zealand 72
nomads 122, 132
Noyon Mountain xii–xiii, 109

Ochirbat Punsalmaa xvi, 16, 31–6, 46,
  54, 85, 138
Odod Gold (company) 74–7, 82
oil supplies 28
Ongi River 85, 90–1, 98
Ongi River Movement (ORM) 81–2,
  86–92, 99, 103, 121, 125–9
Orgilmaa Zundui-Yondon 73–9, 82
original environmentalist society 18
Otgonbayar, S. 65
overgrazing 123, 133
Oyu Tolgoi (OT) mine 40, 57–8, 106,
  137–8
Oyun Sanjaasuren 36

pastoralism 19, 49, 116, 122–33;
  'return' to 124–30; sustainability
  of 123–4, 129–33
Perreault, Tom 2, 41–2
Philippines, the xvi
Polanyi, Karl 139
political leaders in Mongolia 21–2,
  26, 37
political parties in Mongolia 84, 91,
  112; *see also* Motherland Party
populism 139
power relations 83–4

INDEX **157**

precarity, forms of 21–5, 31, 36–7
privatisation 30
'privileged objects' 48
prohibition by law of 'mineral
    exploration and mining operations at
    headwaters of rivers, protected zones
    of the water reservoir and forest area'
    69, 75–6, 82, 96, 104–5, 108–10,
    115, 119, 136
'protective culture' 42–3
protestors 17–18; power of 14–15
Puraam (company) 108

Qian, Sima 49

Raash, R. 91
Rajak, Dinah 13, 63
Reeves, Jeffrey 2
rehabilitation of land 68–74, 77, 93
reification of the national economy
    11–13, 21–5, 51, 58
resistance against mining 73–5, 78
'resource curse' concept 2–3, 13, 15
'resource nationalism' xvi–xvii, 2–4, 7,
    10, 16, 37–8, 40–3, 47–9, 53–4, 58
respect for the state 114
'responsible mining' 96–9
return to the countryside 127–8
Rio Tinto (company) 62, 40, 57–8, 97,
    106, 137–8
river movements 18, 65–6, 92–3, 96–9
rivers, protection of 103–5
Rossabi, Morris 7, 30, 36
Roth, Pedro xiv
Roy, Ravi 6
rule of law 103
Russia 21, 24–8, 31–2; see also Soviet
    Union

Sahlin, Marshall 122
Saikhanbileg Chimed xii–xiv, 58, 137
San Juan, Epifanio 12
Sanchir, Jargalsaikhan 2, 41–2
'scissors' metaphor xiii–xiv
self-funding 99
shamanism 131–2
small- and medium-scale enterprises
    57, 62, 78
Smith, Jane 95
Sneath, David 26, 48, 52–3, 116,
    121, 123
Snow, Keith Harmon 96–7, 113, 115
social movements 102
Sodbaatar, Ya 51
South Sudan 23
sovereignty 6, 23–4, 37
Soviet Union 8–10, 46; see also
    Russia
state-crafting 7–9, 17, 60, 139–41

statism 13, 60–4, 70, 102, 139–40
Steger, Manfred 6
suing the government 110
sustainability 123–4, 129–33

Tavan Tolgoi mine 108
technocrats 43–4, 55
television documentaries 105
Terenguto, A. 124
terrorism 113–14, 137
'third neighbour' policy 21–6,
    29–32, 37
Tolson, Michael 114–15
transnational corporations (TNCs)
    13–15, 83
'trusting relationships' xv–xvi, 53–4
Tseren, P. 124
Tsetsentsolmon, B. 122
Tsing, Anna 83, 94–7, 133
Tsog, L. 107
Turquoise Hill Resources 40

Ulaan Lake 85
Ulaanbaatar 87, 89, 126–9
ündesnii üzel 23, 37, 46
United Movement of Mongolian
    Rivers and Lakes (UMMRL) 96,
    106–7, 109
United Nations 5
United States 24–31; Agency for
    International Development 86
Upton, Caroline 124, 127–8
Uradyn Bulag 27
uranium mining 46
Uuls Zaamar mine 77–9, 82
Uyanga 66, 86
Uyanga Gantomor xii–xiv, 127

Venezuela 101
Verdery, Katherine xv
violence, political 113–14

Walker, Danny 68, 71–9
water supplies 17, 85–6
Watts, Michael 3
Webb, Tristan xv, xviii, 2, 32, 41–3, 53
Welker, Marina 61
Western concepts and practices 99
Weszkalnys, Gisa 2
windfall tax 56–8, 69–70, 74, 82
Winters, Alan xiv
World Bank 26–7, 30–3, 44–5

Xiongnu Empire 48

Zag River 136
zakh zeeliin üye 6
Zandaahüügiin Enhbold 66
Zorig Sanjaasuren 36

CPSIA information can be obtained
at www.ICGtesting.com
Printed in the USA
BVHW041416190920
589199BV00009B/200